MATH MATTERS
for Adults

Whole Numbers WITHDRAWN

♦

Author

Karen Lassiter
Austin Community College
Austin, Texas

♦

Consultants

Connie Eichhorn
Omaha Public Schools
Omaha, Nebraska

M. Gail Joiner Ward
Swainsboro Technical Institute
Swainsboro, Georgia

♦

STECK-VAUGHN
C O M P A N Y
A Subsidiary of National Education Corporation

About the Author

Dr. Karen Lassiter is currently a mathematics instructor at Austin Community College. She is a former Senior Math Editor for Steck-Vaughn Company, and has done extensive work with standardized test preparation. Dr. Lassiter holds a Ph.D. in Educational Research, Testing, and Instructional Design and a bachelor's degree in mathematics and science education from Florida State University.

About the Consultants

Connie Eichhorn is a supervisor of adult education programs in the Omaha Public Schools. A former mathematics teacher and ABE/GED instructor, she earned an undergraduate degree in math at Iowa State University and is completing a doctoral program in adult education at the University of Nebraska. She conducts workshops on math instruction for teachers of adult basic education.

M. Gail Joiner Ward is presently working at Swainsboro Technical Institute as an adult education instructor. Previously she worked with adult learners in a multimedia lab through the School of Education at Georgia Southern University. She has bachelor's and master's degrees in art and early childhood education from Valdosta State College and Georgia Southern University.

Staff Credits

Executive Editor: Ellen Lehrburger
Design Manager: Pamela Heaney
Illustration Credits: Kristian Gallagher, David Griffin, Alan Klemp, Mike Krone
Photo Credits: Cover: (inset) © Superstock, background) © Photri
p. 9 ©Richard Hutchings/PhotoEdit;
p. 27 ©Park Street; p. 53 ©Hazel Hankin/Stock, Boston;
p. 83 ©Park Street; p. 121 ©Park Street; p. 155 ©W.B. Spunbarg/PhotoEdit.
Cover Design: Pamela Heaney

ISBN 0-8114-3650-0

Copyright © 1993 Steck-Vaughn Company

Printed in the United States of America.
3 4 5 6 7 8 9 CK 98 97 96 95 94

Contents

Unit 1

NUMBER SENSE

Unit 2

ADDING WHOLE NUMBERS

Unit 3

SUBTRACTING WHOLE NUMBERS

Unit 4

MULTIPLYING WHOLE NUMBERS

Unit 5

DIVIDING WHOLE NUMBERS

Unit 6

PUTTING YOUR SKILLS TO WORK

To the Learner

The four books in the Steck-Vaughn series *Math Matters for Adults* are *Whole Numbers; Fractions; Decimals and Percents;* and *Measurement, Geometry, and Algebra.* They are written to help you understand and practice arithmetic skills, real-life applications, and problem-solving techniques.

This book contains features which will make it easier for you to work with whole numbers and to apply them to your daily life.

A Skills Inventory test appears at the beginning and end of the book.
•The first test shows you how much you already know.
•The final test can show you how much you have learned.

Each unit has several Mixed Reviews and a Unit Review.
•The Mixed Reviews give you a chance to practice the skills you have learned.
•The Unit Review helps you decide if you have mastered those skills.

There is also a glossary at the end of the book.
•Turn to the glossary to find the meanings of words that are new to you.
•Use the definitions and examples to help strengthen your understanding of terms used in mathematics.

The book contains answers and explanations for the problems.
•The answers let you check your work.
•The explanations take you through the steps used to solve the problems.

Whole Numbers Skills Inventory

Write each number in words.

1. 25 _____ 2. 177 _____

3. 3,511 _____

Compare each set of numbers. Write > or <.

4.
60 ☐ 70

5.
25 ☐ 15

6.
41 ☐ 14

7.
89 ☐ 100

Write the value of the underlined digit in each number.

8.
1̲2 _____

9.
2̲7̲ _____

10.
1̲42 _____

11.
2̲,390 _____

Round each number to the nearest ten.

12.
62 _____

13.
671 _____

14.
1,355 _____

Round each number to the nearest hundred.

15.
417 _____

16.
7,874 _____

17.
33,765 _____

Round each number to the nearest thousand.

18.
8,650 _____

19.
45,099 _____

20.
234,508 _____

Add.

21.
$$\begin{array}{r} 5 \\ +\ 6 \\ \hline \end{array}$$

22.
$3 + 9 + 8 =$

23.
$$\begin{array}{r} 37 \\ +\ 40 \\ \hline \end{array}$$

24.
$8 + 71 + 20 =$

25.
$$\begin{array}{r} 330 \\ +\ 345 \\ \hline \end{array}$$

26.
$$\begin{array}{r} 5,311 \\ 660 \\ +\ \ \ 28 \\ \hline \end{array}$$

27.
$$\begin{array}{r} 34 \\ +\ 57 \\ \hline \end{array}$$

28.
$76 + 83 + 10 =$

29.
$$\begin{array}{r} 547 \\ +\ 367 \\ \hline \end{array}$$

30.
$$\begin{array}{r} 923 \\ 5,687 \\ 15 \\ +\ 1,236 \\ \hline \end{array}$$

31.
$234,000 + 46,677 =$

32.
$$\begin{array}{r} 2,430,000 \\ 3,899 \\ +\ \ \ 24,801 \\ \hline \end{array}$$

Subtract.

33.
$$\begin{array}{r} 13 \\ -\ 6 \\ \hline \end{array}$$

34. $11 - 3 =$

35.
$$\begin{array}{r} 56 \\ -32 \\ \hline \end{array}$$

36.
$$\begin{array}{r} 79 \\ -\ 8 \\ \hline \end{array}$$

37. $38 - 32 =$

38.
$$\begin{array}{r} 699 \\ -\ 37 \\ \hline \end{array}$$

39.
$$\begin{array}{r} 8{,}427 \\ -\ 307 \\ \hline \end{array}$$

40. $9{,}278 - 6{,}152 =$

41.
$$\begin{array}{r} 37 \\ -28 \\ \hline \end{array}$$

42. $82 - 9 =$

43.
$$\begin{array}{r} 40 \\ -\ 7 \\ \hline \end{array}$$

44.
$$\begin{array}{r} 30 \\ -15 \\ \hline \end{array}$$

45.
$$\begin{array}{r} 843 \\ -695 \\ \hline \end{array}$$

46. $726 - 259 =$

47. $5{,}300 - 1{,}486 =$

48.
$$\begin{array}{r} 12{,}020 \\ -\ 1{,}942 \\ \hline \end{array}$$

49.
$$\begin{array}{r} 130{,}000 \\ -\ 99{,}469 \\ \hline \end{array}$$

Multiply.

50. $8 \times 3 =$

51.
$$\begin{array}{r} 52 \\ \times\ 2 \\ \hline \end{array}$$

52. $902 \times 3 =$

53.
$$\begin{array}{r} 3{,}020 \\ \times\ 4 \\ \hline \end{array}$$

54.
$$\begin{array}{r} 23 \\ \times 13 \\ \hline \end{array}$$

55. $530 \times 32 =$

56.
$$\begin{array}{r} 110 \\ \times 123 \\ \hline \end{array}$$

57.
$$\begin{array}{r} 502 \\ \times 438 \\ \hline \end{array}$$

58.
$$\begin{array}{r} 6{,}100 \\ \times\ 100 \\ \hline \end{array}$$

59. $2{,}409 \times 1{,}000 =$

60. $25 \times 7 =$

61.
$$\begin{array}{r} 389 \\ \times\ 4 \\ \hline \end{array}$$

62.
$$\begin{array}{r} 3{,}509 \\ \times\ 5 \\ \hline \end{array}$$

63. $15 \times 38 =$

64. $6{,}070 \times 46 =$

65.
$$\begin{array}{r} 297 \\ \times 163 \\ \hline \end{array}$$

66.
$$\begin{array}{r} 509 \\ \times 600 \\ \hline \end{array}$$

67.
$$\begin{array}{r} 1{,}270 \\ \times\ 302 \\ \hline \end{array}$$

Divide.

68.
$12 \div 6 =$

69.
$5\overline{)30}$

70.
$497 \div 7 =$

71.
$9\overline{)819}$

72.
$8\overline{)59}$

73.
$16 \div 3 =$

74.
$483 \div 5 =$

75.
$4\overline{)1,314}$

76.
$6\overline{)1,303}$

77.
$2\overline{)10,230}$

78.
$86\overline{)93}$

79.
$54\overline{)504}$

80.
$1,309 \div 17 =$

81.
$30\overline{)20,650}$

82.
$832\overline{)8,016}$

83.
$523\overline{)53,870}$

Below is a list of the problems in this Skills Inventory and the pages on which the skills are taught. If you missed any problems, turn to the pages listed and practice the skills. Then correct the problems you missed in the Skills Inventory.

Problem	Practice Page	Problem	Practice Page	Problem	Practice Page
Unit 1		*Unit 3*		58-59	101-102
1-3	10	33-34	55-58	60-61	105-106
4-7	13	35-37	60-61	62	107-108
8-11	18-19	38-40	62	63-64	110-111
12-14	20	41-42	66-67	65	112-113
15-17	21	43-44	68-69	66-67	114-115
18-20	22	45-46	72-73	*Unit 5*	
Unit 2		47-49	75-78	68-69	123-126
21	29-31	*Unit 4*		70-71	127-128
22	32	50	85-87	72-74	129-130
23-24	34-35	51	89-90	75	134-135
25	36	52-53	91	76-77	136-137
26	37	54	95	78-80	140-142
27-28	41-42	55	96	81	143
29	43	56-57	97-98	82	147-148
30	44			83	149-150
31	45				
32	48				

Numbers are everywhere. They are found on road signs, paychecks, gas pumps, book pages, and in advertisements. Number sense can help you find a better buy at the store, understand your paycheck, and figure out how many miles it is from one place to another.

In this unit, you will learn about numbers — what they mean, how they are formed, and how to compare them. You will also learn some of the language used in mathematics.

Getting Ready

You should be familiar with the skills on this page and the next before you begin this unit. To check your answers, turn to page 173.

A number line shows the order of whole numbers. As you move to the right along the number line, the numbers get larger.

Each mark on this number line represents a whole number.

Write the number that each letter represents.

1. A __7__   2. B _____   3. C _____   4. D _____   5. E _____

Getting Ready

 Numbers can be written as words.

Write each number in words. You can use the list below to check your spelling.

0	zero	1	
2		3	
4		5	
6		7	
8		9	
10		11	
12		13	
14		15	
16		17	
18		19	
20		21	twenty-one
22		23	
24		25	
30		40	
50		60	
70		80	
90		100	
1,000			

eight	five	ninety	seventy	thirty	twenty-one
eighteen	forty	one	six	three	twenty-three
eighty	four	one hundred	sixteen	twelve	twenty-two
eleven	fourteen	one thousand	sixty	twenty	two
fifteen	nine	seven	ten	twenty-five	zero
fifty	nineteen	seventeen	thirteen	twenty-four	

Number Patterns

You can create number patterns by counting forward or backward by a certain number. For example, when you count forward by 1, you create a pattern. Each number is one more than the number before it. To find the next number in the pattern below, count 1 more.

4, 5, 6, __7__
rule: count 1 more

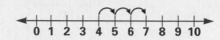

When you count backward by 3, you also create a pattern. Each number is 3 less than the number before it. To find the next number in the pattern below, count 3 less.

10, 7, 4, __1__
rule: count 3 less

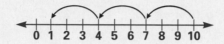

Use These Steps

Find the rule for the pattern 6, 4, 2, _____. Then write the missing number.

1. Look at each number in the pattern 6, 4, 2.

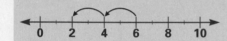

2. Write the rule for the pattern.

rule: **count 2 less**

3. Use the rule to find the missing number.

6, 4, 2, __0__

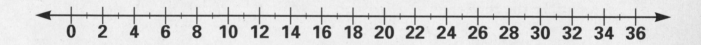

Find the rule for each pattern. Then write the missing number.

1. 5, 8, 11, __14__
rule: __count 3 more__

2. 5, 10, 15, _____
rule: _counts more._

3. 8, 12, 16, _____
rule: _____

4. 13, 11, 9, __7__
rule: __count 2 less__

5. 9, 6, 3, _____
rule: _count less_

6. 17, 14, 11, _____
rule: _____

7. 18, 22, 26, _____
rule: _count 4 more_

8. 27, 25, 23, _____
rule: _____

9. 33, 26, 19, _____
rule: _____

10. 4, 10, 16, _____
rule: _coun_

11. 24, 19, 14, _____
rule: _____

12. 10, 7, 4, _____
rule: _____

Number Groups

Our number system is based on tens. We can write two-digit numbers as groups of tens and ones. We can write three-digit numbers as groups of hundreds, tens, and ones.

  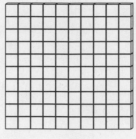 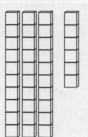

5 ones = 5   2 tens 3 ones = 23   1 hundred   3 tens   6 ones = 136

Use These Steps

Write 104 as groups of hundreds, tens, and ones.

1. Look at the digit on the left. Write the number of hundreds that this digit shows.

    104

    _1_ hundreds

2. Look at the digit in the center. Write the number of tens that this digit shows.

    104

    _0_ tens

3. Look at the digit on the right. Write the number of ones that this digit shows.

    104

    _4_ ones

Write each digit in the correct group.

1. 2 _2_ ones

2. 6 ___ ones

3. 15 _1_ tens _5_ ones

4. 89 ___ tens ___ ones

5. 10 ___ tens ___ ones

6. 20 ___ tens ___ ones

7. 50 ___ tens ___ ones

8. 90 ___ tens ___ ones

9. 254 _2_ hundreds _5_ tens _4_ ones

10. 971 ___ hundreds ___ tens ___ ones

11. 855 ___ hundreds ___ tens ___ ones

12. 482 ___ hundreds ___ tens ___ ones

13. 106 ___ hundreds ___ tens ___ ones

14. 605 ___ hundreds ___ tens ___ ones

15. 408 ___ hundreds ___ tens ___ ones

16. 302 ___ hundreds ___ tens ___ ones

17. 500 ___ hundreds ___ tens ___ ones

18. 100 ___ hundreds ___ tens ___ ones

19. 700 ___ hundreds ___ tens ___ ones

20. 600 ___ hundreds ___ tens ___ ones

Comparing Numbers

A number line can be used to compare numbers. Remember, as you move to the right, the numbers get larger.

This number line shows whole numbers from 10 to 40.

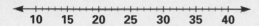

The symbol > means *greater than*.   25 > 20
The symbol < means *less than*.   15 < 20

Use These Steps

Compare 23 and 32.

1. Find 23 and 32 on this part of the number line.

2. Decide which number is farther right. It is the greater number. The number that is farther left is the lesser number.

3. Use the symbols to show *greater than* or *less than*.

32 is greater than 23
or
23 is less than 32

32 > 23
or
23 < 32

Compare each set of numbers. Write > or <.

1. 50 $\boxed{>}$ 40   2. 10 $\boxed{<}$ 20   3. 80 ☐ 10   4. 70 ☐ 30

5. 25 ☐ 30   6. 51 ☐ 40   7. 10 ☐ 29   8. 30 ☐ 42

9. 100 ☐ 96   10. 52 ☐ 34   11. 48 ☐ 54   12. 64 ☐ 68

13. 79 ☐ 82   14. 63 ☐ 83   15. 91 ☐ 81   16. 22 ☐ 33

17. 41 ☐ 22   18. 33 ☐ 35   19. 100 ☐ 10   20. 89 ☐ 99

21. 42 ☐ 24   22. 89 ☐ 98   23. 19 ☐ 91   24. 0 ☐ 10

Number Meaning

Every number is made up of other numbers. You can use several combinations of numbers to express a given number. For example, to pay for a shirt that costs $12, you could use 1 five-dollar bill and 7 one-dollar bills. You could also use 1 ten-dollar bill and 2 one-dollar bills. The number line below shows both combinations.

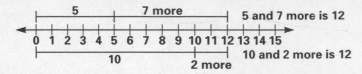

You could also express 12 by counting backwards from a number greater than 12. 15 and 3 less is 12.

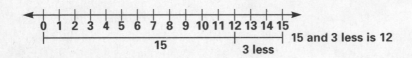

Use These Steps

Write the number 9 two different ways.

1. Choose a number less than 9. Try 5. Count forward to 9 from 5.

2. Choose a number greater than 9. Try 10. Count backward to 9 from 10.

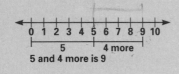
5 and 4 more is 9

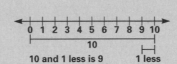
10 and 1 less is 9

Use the number line to write each number two different ways.

1. 5

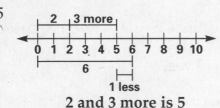

2 and 3 more is 5

6 and 1 less is 5

2. 8

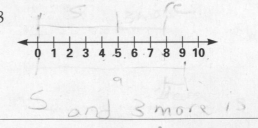

S and 3 more is 8

9 and 1 less is 8

3. 10

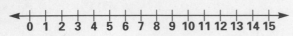

4. 14

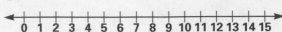

Real-Life Application Daily Living

You use numbers when you measure things. For example, you measure weight in numbers of ounces and pounds, and cost in numbers of dollars and cents.

The groups below show measurements we use every day. The sign = means *equals*.

12 inches	= 1 foot
3 feet	= 1 yard
5,280 feet	= 1 mile

16 ounces	= 1 pound
2,000 pounds	= 1 ton

100 cents	= 1 dollar

60 seconds	= 1 minute
60 minutes	= 1 hour
24 hours	= 1 day
365 days	= 1 year

Circle the letter of the most likely measurement.

1. the weight of a chair
    a. 18 pounds
    b. 18 ounces

2. the weight of a can of soup
    a. 10 pounds
    b. 10 ounces

3. the height of a telephone pole
    a. 30 feet
    b. 30 inches

4. the time it takes to brush your teeth
    a. 1 minute
    b. 1 second

5. the length of a room
    a. 12 inches
    b. 12 feet

6. the time it takes to paint a house
    a. 3 days
    b. 3 years

7. the time it takes to watch a movie
    a. 2 days
    b. 2 hours

8. the cost of a piece of bubble gum
    a. 10 cents
    b. 10 dollars

9. the weight of an elephant
    a. 2 tons
    b. 2 pounds

10. the length of a football field
    a. 100 yards
    b. 100 miles

11. the cost of a movie ticket
    a. 6 cents
    b. 6 dollars

12. the time it takes to eat dinner
    a. 20 hours
    b. 20 minutes

Mixed Review

Find the rule for each pattern. Then write the missing number.

(number line: 0 5 10 15 20 25 30 35 40 45 50 55)

1. 8, 6, 4, _____

   rule: _____

2. 18, 15, 12, _____

   rule: _____

3. 11, 16, 21, _____

   rule: _____

4. 7, 14, 21, _____

   rule: _____

5. 24, 18, 12, _____

   rule: _____

6. 51, 44, 37, _____

   rule: _____

Write each digit in the correct group.

7. 4 ____ ones

8. 28 ____ tens ____ ones

9. 134 ____ hundreds ____ tens ____ ones

10. 614 ____ hundreds ____ tens ____ ones

Compare each set of numbers. Write > or <.

11. 25 _____ 31

12. 42 _____ 22

13. 37 _____ 41

14. 20 _____ 50

15. 44 _____ 39

16. 51 _____ 15

Use the number lines to write each number two different ways.

17. 4

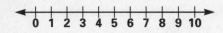

(number line: 0 1 2 3 4 5 6 7 8 9 10)

_____

_____

18. 7

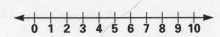

(number line: 0 1 2 3 4 5 6 7 8 9 10)

_____

_____

19. 11

(number line: 0 1 2 3 4 5 6 7 8 9 10 11 12 13 14 15)

_____

_____

20. 13

(number line: 0 1 2 3 4 5 6 7 8 9 10 11 12 13 14 15)

_____

_____

Circle the letter of the most likely measurement.

21. the weight of an apple
    a. 10 pounds
    b. 10 ounces

22. the cost of a car
    a. ten thousand dollars
    b. ten dollars

Real-Life Application

The Lee family goes to Central Park to ride their bicycles three times each week. They are practicing for a race.

Solve.

1. John Lee rode his bike 36 miles last week and 32 miles this week. Did he ride more miles last week or this week?

   Answer _____

2. Rosa Lee rode her bike 34 miles last week and 43 miles this week. Did she ride more miles last week or this week?

   Answer _____

3. John rode a total of 68 miles for the two weeks. Rosa rode a total of 77 miles. Who rode the greater distance: John or Rosa?

   Answer _____

4. Rosa's sister Cecilia rides bikes with John and Rosa. Cecilia is 25 years old. Rosa is 29 years old. Who is older: Rosa or Cecilia?

   Answer _____

5. Last Sunday the temperature in the park reached 102°. This Sunday, the temperature reached 92°. On which Sunday was the temperature greater than 100°: last Sunday or this Sunday?

   Answer _____

6. When was the temperature in Central Park less than 100°: last Sunday or this Sunday?

   Answer _____

7. The weather station measured 52 inches of rainfall in Central Park for this year. The record rainfall for Central Park is 50 inches. Is this year's rainfall greater or less than the record?

   Answer _____

8. Last year 300 riders entered the race. Two years ago 250 riders entered the race. Were there more riders last year or 2 years ago?

   Answer _____

Place Value to the Hundreds Place

The value of a digit in a number depends on its place. For example, you know that 25 is a group of 2 tens and 5 ones. In 25, the value of the 2 is 20, and the value of the 5 is 5.

The place value chart shows the digits in the number 25 in their correct groups.

Use These Steps

Write the value of each digit in the number 590.

1. Write each digit in its correct place in the chart below.

2. Write each digit in the correct number group. Then write the value of the digit.

   5 hundreds = 500

   9 tens   = 90

   0 ones   = 0

Write each number in the chart. Then write each digit in the correct number group, and write the value of the digit.

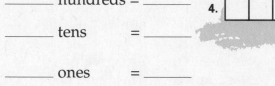

1. 67

   __6__ tens = __60__

   __7__ ones = __7__

2. 92

   _____ tens = _____

   _____ ones = _____

3. 125

   __1__ hundreds = __100__

   __2__ tens   = __20__

   __5__ ones   = __5__

4. 490

   _____ hundreds = _____

   _____ tens   = _____

   _____ ones   = _____

Write the value of the underlined digit in each number.

5. 5<u>7</u> __50__   6. 3<u>6</u> _____   7. <u>9</u>8 _____   8. <u>1</u>35 _____

9. 3<u>7</u>9_____   10. 85<u>4</u> _____   11. <u>2</u>05 _____   12. <u>3</u>00 _____

13. 5<u>7</u>0_____   14. 49<u>9</u> _____   15. 6<u>8</u>7 _____   16. 7<u>0</u>4 _____

Place Value to the Thousands Place

A place value chart can also show thousands. The number 4,189 is a group of 4 thousands, 1 hundred, 8 tens, and 9 ones.

Commas are used to separate digits into groups of three. This makes large numbers easier to read.

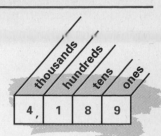

Use These Steps

Write the value of each digit in the number 7,036.

1. Write each digit in its correct place in the chart below.

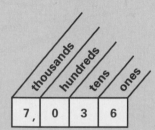

2. Write each digit beside the correct number group. Then write the value of the digit.

__7__ thousands = __7,000__

__0__ hundreds = __000__

__3__ tens = __30__

__6__ ones = __6__

Write each number in the chart. Then write each digit in the correct number group, and write the value of the digit.

1. 2,437

    __2__ thousands = __2,000__

    __4__ hundreds = __400__

    __3__ tens = __30__

    __7__ ones = __7__

2. 3,013

    _____ thousands = _____

    _____ hundreds = _____

    _____ tens = _____

    _____ ones = _____

	thousands	hundreds	tens	ones
1.	2,	4	3	7
2.				
3.				
4.				

3. 4,608

    _____ thousands = _____

    _____ hundreds = _____

    _____ tens = _____

    _____ ones = _____

4. 5,520

    _____ thousands = _____

    _____ hundreds = _____

    _____ tens = _____

    _____ ones = _____

Write the value of the underlined digit in each number.

5. 5̲67   __500__

6. 69̲0 _____

7. 70̲3̲ _____

8. 1,33̲4̲ _____

9. 3̲,701 _____

10. 8̲,006 _____

11. 6,4̲00 _____

12. 5,632̲ _____

13. 9,95̲5 _____

Rounding to the Nearest Ten

In working with numbers, sometimes you do not need to use an exact number. Instead, you need to know *about how many*, rather than *exactly how many*. To find *about how many*, use rounding.

You can use a number line to round a given number. To round a number to the nearest ten, find a number in the tens place that is closer to that number. That is the number you round to.

If the digit in the ones place in the number you're rounding is less than 5, round down to the nearest ten. If the digit in the ones place is 5 (is halfway between two tens) or is greater than 5, round up to the higher number in the tens place.

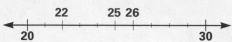

22 is closer to 20 than to 30   22 rounds down to 20
25 is halfway between 20 and 30   25 rounds up to 30
26 is closer to 30 than to 20   26 rounds up to 30

You can round a number without using a number line. Follow the steps below.

Use These Steps

Round 37 to the nearest ten.

1. Look at the digit in the ones place. Is it less than 5, exactly 5, or greater than 5?

   3<u>7</u>

   7 > 5

2. Since the digit in the ones place is greater than 5, round up to the nearest ten.

   37 rounds up to 40

Round each number to the nearest ten.

1. 45 ___50___

2. 58 _____

3. 54 _____

4. 52 _____

5. 71 _____

6. 55 _____

7. 886 ___890___

8. 911 _____

9. 448 _____

10. 794 _____

11. 385 _____

12. 176 _____

13. 9,554 ___9,550___

14. 1,406 _____

15. 6,501 _____

16. 2,004 _____

17. 8,765 _____

18. 1,979 _____

19. 44 _____

20. 224 _____

21. 903 _____

22. 1,889 _____

23. 75 _____

24. 9,989 _____

Rounding to the Nearest Hundred

To round a number to the nearest hundred, find a number in the hundreds place that is closer to that number. That is the number you round to.

If the digit in the tens place in the number you're rounding is less than 5, round down to the nearest hundred. If the digit in the tens place is 5 or is greater than 5, round up to the higher number in the hundreds place.

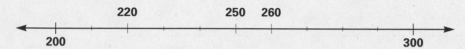

220 is closer to 200 than to 300   220 rounds down to 200
250 is halfway between 200 and 300   250 rounds up to 300
260 is closer to 300 than to 200   250 rounds up to 300

You can round a number without using a number line.
Follow the steps below.

Use These Steps

Round 370 to the nearest hundred.

1. Look at the digit in the tens place. Is it less than 5, exactly 5, or greater than 5?

   3 7 0

   7 > 5

2. Since the digit in the tens place is greater than 5, round up to the nearest hundred.

   370 rounds up to 400

Round each number to the nearest hundred.

1.	410	<u>400</u>	**2.**	540	_____	**3.**	650 _____
4.	649	_____	**5.**	733	_____	**6.**	421 _____
7.	1,550	<u>1,600</u>	**8.**	2,640	_____	**9.**	4,680 _____
10.	1,242	_____	**11.**	1,101	_____	**12.**	2,142 _____
13.	16,660	<u>16,700</u>	**14.**	24,441	_____	**15.**	92,621 _____
16.	18,492	_____	**17.**	10,401	_____	**18.**	13,899 _____
19.	780	_____	**20.**	50,149	_____	**21.**	9,790 _____
22.	2,560	_____	**23.**	33,842	_____	**24.**	569 _____

Rounding to the Nearest Thousand

To round a number to the nearest thousand, find a number in the thousands place that is closer to that number. That is the number you round to.

If the digit in the hundreds place in the number you're rounding is less than 5, round down to the nearest thousand. If the digit in the hundreds place is 5 or is greater than 5, round up to the higher number in the thousands place.

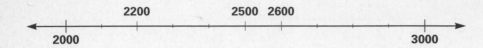

2,200 is closer to 2,000 than to 3,000   2,200 rounds down to 2,000
2,500 is halfway between 2,000 and 3,000   2,500 rounds up to 3,000
2,600 is closer to 3,000 than to 2,000   2,600 rounds up to 3,000

You can round a number without using a number line.
Follow the steps below.

Use These Steps

Round 3,700 to the nearest thousand.

1. Look at the digit in the hundreds place. Is it less than 5, exactly 5, or greater than 5?

   3, 7 00

   7 > 5

2. Since the digit in the hundreds place is greater than 5, round up to the nearest thousand.

   3,700 rounds up to 4,000

Round each number to the nearest thousand.

1.	7,300	7,000	2.	5,500	_____	3.	4,100	_____
4.	4,490	_____	5.	6,210	_____	6.	4,530	_____
7.	18,081	18,000	8.	97,999	_____	9.	36,561	_____
10.	21,801	_____	11.	13,569	_____	12.	92,489	_____
13.	111,591	112,000	14.	213,649	_____	15.	139,492	_____
16.	264,949	_____	17.	955,611	_____	18.	231,499	_____
19.	8,700	_____	20.	149,300	_____	21.	47,499	_____
22.	561,180	_____	23.	86,399	_____	24.	6,790	_____

Problem Solving: Using a Table

The table to the right shows the 1980 and 1990 populations of five cities. Some of these cities had more people in 1990 than they had in 1980. Some of them had fewer people. Use the table to answer the questions below.

	1980	1990
Pattonville	12,965	13,012
Shoreline	23,312	23,501
Eagle City	11,573	11,416
Benton	24,599	24,467
Hillview	22,207	22,299

Population Figures for Five Cities

Example   Which city had the largest population in 1980?

▶ **Step 1.** Look at the populations in the column for 1980.

▶ **Step 2.** Compare the values of the digits on the left in each number. If the digits are the same, compare the next digits.

<u>12</u>,965
<u>23</u>,312
<u>11</u>,573
<u>24</u>,599
<u>22</u>,207

▶ **Step 3.** Find the largest number. Write the name of the city that had the largest population in 1980.

Benton

Solve.

1. Which city had the smallest population in 1980?

2. Which cities had more people in 1990 than in 1980?

Answer _____

Answer _____

3. Which cities had fewer people in 1990 than in 1980?

4. Which city had the larger population in 1990: Pattonville or Shoreline?

Answer _____

Answer _____

5. Which city had the smaller population in 1990: Eagle City or Pattonville?

Answer _____

6. Which city had the largest population in 1990?

Answer _____

7. Was the population of Benton smaller in 1980 or in 1990?

Answer _____

8. Which city had the smallest population in 1990?

Answer _____

9. Which city had the smaller population in 1980: Hillview or Shoreline?

Answer _____

10. Was the population of Shoreline larger in 1990 or in 1980?

Answer _____

11. Write the name and population of each city in the table below. Then for each year, round the populations to the nearest thousand.

City	1980 Population	Nearest Thousand	1990 Population	Nearest Thousand

Unit 1 *Review*

Write each number in words.

1. 25 _____

2. 41 _____

3. 987 _____

4. 603 _____

5. 1,001 _____

6. 6,852 _____

Find the rule for each pattern. Then write the missing number.

7. 35, 45, 55, _____   8. 18, 20, 22, _____   9. 3, 6, 9, _____

   rule: _____   rule: _____   rule: _____

10. 15, 13, 11, _____   11. 20, 30, 40, _____   12. 14, 12, 10, _____

    rule: _____   rule: _____   rule: _____

Compare each set of numbers. Write > or <.

13. 20 ☐ 19   14. 44 ☐ 72   15. 11 ☐ 10   16. 38 ☐ 98

17. 90 ☐ 70   18. 64 ☐ 40   19. 44 ☐ 33   20. 22 ☐ 26

Use the number lines to write each number two different ways.

21. 6

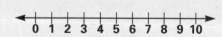

_____

_____

22. 2

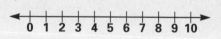

_____

_____

23. 12

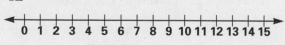

_____

_____

24. 9

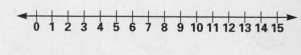

_____

_____

Circle the letter of the most likely measurement.

25. the length of a baseball bat
    a. 3 feet
    b. 3 yards

26. the height of a flagpole
    a. 20 miles
    b. 20 feet

27. the weight of a pair of glasses
    a. 1 pound
    b. 1 ounce

28. the cost of a can of soda
    a. 60 cents
    b. 60 dollars

Write the value of the underlined digit in each number.

29. 1̲73 _____   30. 4,52̲2 _____   31. 9̲07 _____

32. 8̲,406 _____   33. 5,61̲7 _____   34. 9,24̲0 _____

Round each number to the nearest ten.

35. 84 _____   36. 15 _____   37. 59 _____   38. 104 _____

39. 155 _____   40. 271 _____   41. 1,235 _____   42. 4,556 _____

Round each number to the nearest hundred.

43. 456 _____   44. 923 _____   45. 555 _____   46. 199 _____

47. 1,222 _____   48. 5,889 _____   49. 33,499 _____   50. 58,541 _____

Round each number to the nearest thousand.

51. 7,456 _____   52. 4,499 _____   53. 9,489 _____   54. 10,732 _____

55. 34,567 _____   56. 99,099 _____   57. 167,998 _____   58. 132,721 _____

Below is a list of the problems in this review and the pages on which the skills are taught. If you missed any problems, turn to the pages listed and practice the skills. Then correct the problems you missed in the Unit Review.

Unit 2 ADDING WHOLE NUMBERS

You add whole numbers when you count the number of hours you work in a week, the number of calories you eat in a day, or the number of miles you travel from one place to another.

In this unit, you will learn the addition facts and how to line up and add whole numbers. You will also learn how to estimate to find an approximate answer when an exact answer is not needed.

Getting Ready

You should be familiar with the skills on this page and the next before you begin this unit. To check your answers, turn to page 176.

 When you write out numbers, you need to know the place value of each digit.

number	place value
25	2 tens 5 ones
392	3 hundreds 9 tens 2 ones
1,200	1 thousand 2 hundreds 0 tens 0 ones

Write each digit in the correct group.

1. 37 ___**3**___ tens ___**7**___ ones

2. 550 _____ hundreds _____ tens _____ ones

3. 3,901_____ thousands _____ hundreds _____ tens _____ ones

For review, see Unit 1, page 12.   **27**

Getting Ready

 When you add whole numbers, you need to know the place value of each digit. The place value chart shows the value of each digit.

3,476

13

940

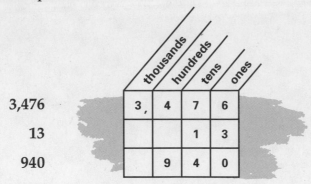

Write the value of the underlined digit in each number.

4. 2,1<u>6</u>1 **2,000**   5. 2<u>5</u> _____   6. 6<u>4</u>3 _____   7. <u>8</u>17 _____   8. 4,<u>2</u>00 _____

9. <u>5</u>6 _____   10. 50<u>9</u> _____   11. 10<u>2</u> _____   12. <u>9</u>,094 _____   13. 7<u>8</u> _____

For review, see Unit 1, pages 18-19.

 When you round whole numbers, you need to know the place value of each digit.

Round to the nearest ten: 15 rounds to   20
Round to the nearest hundred: 203 rounds to   200
Round to the nearest thousand: 7,960 rounds to 8,000

Round each number to the nearest ten.

14. 35 _____ **40** _____   15. 122 _____   16. 309 _____   17. 2,449 _____

Round each number to the nearest hundred.

18. 135 _____ **100** _____   19. 4,559 _____   20. 16,630 _____   21. 89 _____

Round each number to the nearest thousand.

22. 1,087 _____ **1,000** _____   23. 15,987 _____   24. 234,191 _____   25. 999 _____

For review, see Unit 1, pages 20-22.

Addition Facts

To add larger numbers, you should first know the basic addition facts.
You will find it helpful to know the following facts by heart.

Add the following numbers to complete each row. Notice that the answers form a pattern.

1.
$$\begin{array}{cccccccccc} 0 & 0 & 0 & 0 & 0 & 0 & 0 & 0 & 0 & 0 \\ +0 & +1 & +2 & +3 & +4 & +5 & +6 & +7 & +8 & +9 \\ \hline 0 & 1 & 2 & 3 & 4 & 5 & 6 & 7 & 8 & 9 \end{array}$$

2.
$$\begin{array}{cccccccccc} 1 & 1 & 1 & 1 & 1 & 1 & 1 & 1 & 1 & 1 \\ +0 & +1 & +2 & +3 & +4 & +5 & +6 & +7 & +8 & +9 \\ \hline \end{array}$$

3.
$$\begin{array}{cccccccccc} 2 & 2 & 2 & 2 & 2 & 2 & 2 & 2 & 2 & 2 \\ +0 & +1 & +2 & +3 & +4 & +5 & +6 & +7 & +8 & +9 \\ \hline \end{array}$$

4.
$$\begin{array}{cccccccccc} 3 & 3 & 3 & 3 & 3 & 3 & 3 & 3 & 3 & 3 \\ +0 & +1 & +2 & +3 & +4 & +5 & +6 & +7 & +8 & +9 \\ \hline \end{array}$$

5.
$$\begin{array}{cccccccccc} 4 & 4 & 4 & 4 & 4 & 4 & 4 & 4 & 4 & 4 \\ +0 & +1 & +2 & +3 & +4 & +5 & +6 & +7 & +8 & +9 \\ \hline \end{array}$$

6.
$$\begin{array}{cccccccccc} 5 & 5 & 5 & 5 & 5 & 5 & 5 & 5 & 5 & 5 \\ +0 & +1 & +2 & +3 & +4 & +5 & +6 & +7 & +8 & +9 \\ \hline \end{array}$$

7.
$$\begin{array}{cccccccccc} 6 & 6 & 6 & 6 & 6 & 6 & 6 & 6 & 6 & 6 \\ +0 & +1 & +2 & +3 & +4 & +5 & +6 & +7 & +8 & +9 \\ \hline \end{array}$$

8.
$$\begin{array}{cccccccccc} 7 & 7 & 7 & 7 & 7 & 7 & 7 & 7 & 7 & 7 \\ +0 & +1 & +2 & +3 & +4 & +5 & +6 & +7 & +8 & +9 \\ \hline \end{array}$$

9.
$$\begin{array}{cccccccccc} 8 & 8 & 8 & 8 & 8 & 8 & 8 & 8 & 8 & 8 \\ +0 & +1 & +2 & +3 & +4 & +5 & +6 & +7 & +8 & +9 \\ \hline \end{array}$$

10.
$$\begin{array}{cccccccccc} 9 & 9 & 9 & 9 & 9 & 9 & 9 & 9 & 9 & 9 \\ +0 & +1 & +2 & +3 & +4 & +5 & +6 & +7 & +8 & +9 \\ \hline \end{array}$$

Addition Facts Practice

Use the addition facts from page 29 to complete the table.

Example   Find an empty box in the table. Look up the column to the top number. Then move left from the box to the number at the beginning of the row. Add these two numbers. Write the answer in the box.

+	0	1	2	3	4	5	6	7	8	9
0	0	1	2	3	4	5				
1	1	2	3	4						
2	2	3	4							
3	3	4								
4	4									
5	5									
6										
7										
8										
9										

To use the table to complete addition facts, find one number at the top of a column, and the other number at the beginning of a row. The sum is the number in the box where the row and column meet.

Complete the following addition facts.

1. $3 + 6 = 9$

2. $3 + 9 =$

3. $4 + 8 =$

4. $0 + 2 =$

5. $6 + 7 =$

6. $9 + 2 =$

7. $6 + 6 =$

8. $8 + 6 =$

9. $1 + 1 =$

10. $7 + 3 =$

11. $4 + 9 =$

12. $5 + 6 =$

13. $2 + 7 =$

14. $5 + 9 =$

15. $8 + 1 =$

16. $3 + 2 =$

17. $9 + 7 =$

18. $5 + 5 =$

19. $9 + 9 =$

20. $6 + 4 =$

Addition Facts Practice

Addition problems can be written vertically or horizontally.

$$\begin{aligned} \text{addend} &\longrightarrow 5 \\ \text{addend} &\longrightarrow \underline{+\ 1} \quad \text{is the same as} \quad 5 + 1 = 6 \\ \text{sum} &\longrightarrow 6 \end{aligned}$$

Fill in the boxes to complete the following addition facts. You can use the table on page 30 if you need help remembering the facts.

1. $8 + 6 = \boxed{14}$

2. $9 + 1 = \square$

3. $7 + 9 = \square$

4. $6 + 3 = \square$

5. $5 + \square = 6$

6. $3 + \square = 5$

7. $4 + \square = 10$

8. $7 + \square = 13$

9. $\square + 2 = 2$

10. $\square + 4 = 7$

11. $\square + 6 = 14$

12. $\square + 9 = 15$

13. $4 + 8 = \square$

14. $\square + 7 = 16$

15. $0 + \square = 0$

16. $8 + 3 = \square$

17. $2 + \square = 8$

18. $7 + 9 = \square$

19. $2 + \square = 3$

20. $0 + \square = 9$

21. $6 + 4 = \square$

22. $1 + \square = 6$

23. $8 + \square = 12$

24. $\square + 2 = 7$

Fill in the boxes with any numbers that make the sums. There may be more than one set of numbers that makes a true statement.

25. $\boxed{5} + \boxed{4} = 9$

26. $\square + \square = 15$

27. $\square + \square = 18$

28. $\square + \square = 12$

29. $\square + \square = 5$

30. $\square + \square = 13$

31. $\square + \square = 7$

32. $\square + \square = 4$

33. $\square + \square = 16$

34. $\square + \square = 10$

35. $\square + \square = 14$

36. $\square + \square = 11$

37. $\square + \square = 17$

38. $\square + \square = 8$

39. $\square + \square = 6$

40. $\square + \square = 3$

Column Addition

You can use the addition facts to add three or more numbers in a column.

Use These Steps

Add 4
    3
  +2

1. Be sure that the digits are lined up in a column.

    4
    3
  +2

2. Add the first two digits.

    4     $4 + 3 = 7$
    3
  +2

3. Add the last digit to the sum of the first two digits.

    4
    3
  +2 $2 + 7 = 9$
    9

Add.

1.
  4
  4
+3
11

2.
  6
  2
+2

3.
  3
  1
+5

4.
  8
  1
+1

5.
  5
  2
+6

6.
  4
  1
+6

7.
$9 + 0 + 2 =$
  9
  0
+2
11

8. $3 + 2 + 1 =$

9. $6 + 3 + 3 =$

10. $4 + 7 + 2 =$

11.
$5 + 6 + 1 + 6 =$
  5
  6
  1
+6
18

12. $7 + 3 + 9 + 2 =$

13. $0 + 5 + 8 + 4 =$

14. $3 + 5 + 7 + 6 =$

15. Roy worked for 4 hours on Friday, 6 hours on Saturday, and 5 hours on Sunday. How many total hours did he work during these three days?

16. Jean's tomato plant grew 3 tomatoes one week, 6 tomatoes the next week, 4 tomatoes the next week, and 5 tomatoes the week after that. How many tomatoes did Jean grow in all?

Answer_____

Answer_____

You can use addition to make decisions when you buy items at the store.

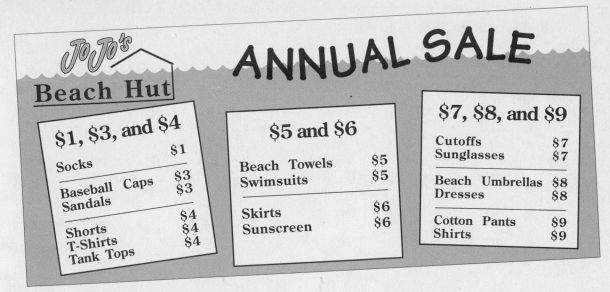

JoJo's Beach Hut — ANNUAL SALE

$1, $3, and $4

Socks	$1
Baseball Caps	$3
Sandals	$3
Shorts	$4
T-Shirts	$4
Tank Tops	$4

$5 and $6

Beach Towels	$5
Swimsuits	$5
Skirts	$6
Sunscreen	$6

$7, $8, and $9

Cutoffs	$7
Sunglasses	$7
Beach Umbrellas	$8
Dresses	$8
Cotton Pants	$9
Shirts	$9

Use the prices in the advertisement to find the answer to each problem. Circle the letter beside the correct answer.

1. Harry wants to buy a beach umbrella and a T-shirt at Jojo's annual sale. He has $15 to spend.

   a. He has exactly the right amount of money.
   b. He does not have enough money.
   c. He will get change back when he buys both items.

2. Dolores has $10 to spend at Jojo's. She could buy a beach towel and

   a. a skirt.
   b. sandals.
   c. cutoffs.

3. If Jasmine buys sunglasses, a beach towel, and sandals, the total bill will be

   a. more than $10.
   b. exactly $10.
   c. less than $10.

4. Carl bought cotton pants and a shirt. Kai bought a baseball cap, a tank top, socks, and a beach umbrella.

   a. Carl spent more money than Kai.
   b. Kai spent more money than Carl.
   c. They spent the same amount of money.

5. Chris has $20 to spend. He wants to buy sunglasses and a shirt. He also has enough money to buy

   a. a beach umbrella.
   b. cutoffs.
   c. sandals.

6. Sue bought a tank top and sunglasses at the sale. For the same amount of money, she could have bought

   a. a dress and shorts.
   b. a skirt and a beach towel.
   c. a swimsuit and a beach umbrella.

Adding Two-Digit Numbers

You can use the addition facts to add two-digit numbers. Just be sure to put your answers in the correct columns. You may want to use the table you completed on page 30 to help you.

Use These Steps

Add   37
       + 21

1. Be sure that the ones and the tens are lined up.

```
  3 7
+ 2 1
```

2. Add the ones. 7 + 1 = 8 ones. Put the 8 ones in the ones column.

```
  37
+ 21
   8
```
← 8 ones

3. Add the tens. 3 + 2 = 5 tens. Put the 5 tens in the tens column.

```
  37
+ 21
  58
```
← 5 tens

Add.

1.	2.	3.	4.	5.	6.
50 + 34 84	65 + 12	41 + 58	36 + 31	72 + 20	33 + 15

7.	8.	9.	10.	11.	12.
90 + 6 96	45 + 3	17 + 1	54 + 4	38 + 1	16 + 2

13.	14.	15.	16.	17.	18.
76 + 23	81 + 10	49 + 40	60 + 4	37 + 1	52 + 15

19.	20.	21.	22.	23.	24.
71 + 16	25 + 4	30 + 5	66 + 22	48 + 30	31 + 14

25.	26.	27.	28.	29.	30.
80 + 1	47 + 22	58 + 21	50 + 8	61 + 34	72 + 5

Adding Two-Digit Numbers

To add numbers that are not lined up, first put the digits in columns.
Line up the digits that have the same place value.

$50 + 25$

```
 tens ones
  5   0
+ 2   5
  7   5
```

Use These Steps

Add 60 + 12 + 4

1. Write the digits in columns so that the digits with the same place values are lined up.

```
  60
  12
+  4
```

2. Add the ones.
$0 + 2 + 4 = 6$ ones.

```
  60
  12
+  4
   6
```
↖ 6 ones

3. Add the tens.
$6 + 1 = 7$ tens.

```
  60
  12
+  4
  76
```
↖ 7 tens

Add.

1.
$42 + 27 =$
```
  42
+ 27
  69
```

2.
$15 + 13 =$

3.
$51 + 44 =$

4.
$70 + 14 =$

5.
```
  56
+ 12
```

6.
```
  33
+ 16
```

7.
```
  50
+ 27
```

8.
```
  34
+ 10
```

9.
```
  25
+ 24
```

10.
$42 + 20 + 16 =$
```
  42
  20
+ 16
  78
```

11.
$32 + 24 + 11 =$

12.
$30 + 12 + 6 =$

13.
$28 + 11 + 30 =$

14. Marta walked 10 miles in one week, 5 miles the next week, and 4 miles the week after that. How many miles did she walk in all?

15. John bought a notebook for $1 and jeans for $12 at Arnie's Discount Store. How much did he spend all together?

Answer_____

Answer_____

Adding Three-Digit Numbers

To add three-digit numbers, first be sure that the digits with the same place values are lined up in the ones, tens, and hundreds columns. Add only the digits that have the same place value. Be sure to put your answers in the correct columns.

Use These Steps

Add   360
     36
  + 3

1. Be sure that the digits are lined up.

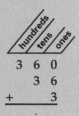

```
3 6 0
  3 6
+   3
```

2. Add each column, starting with the digits in the ones place.

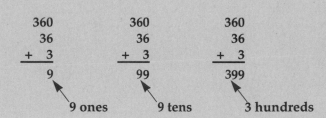

$$
\begin{array}{r} 360 \\ 36 \\ +\ 3 \\ \hline 9 \end{array}
\qquad
\begin{array}{r} 360 \\ 36 \\ +\ 3 \\ \hline 99 \end{array}
\qquad
\begin{array}{r} 360 \\ 36 \\ +\ 3 \\ \hline 399 \end{array}
$$

9 ones   9 tens   3 hundreds

Add.

1.
```
  106
+  23
  129
```

2.
```
  794
+ 105
```

3.
```
   32
+ 142
```

4.
```
  181
+ 816
```

5.
```
  431
+   4
```

6.
```
  670
+ 129
```

7.
```
   36
+ 502
```

8.
```
    1
+ 378
```

9.
```
  322
+  76
```

10.
```
  210
+   8
```

11.
```
  203
  222
+ 164
  589
```

12.
```
  622
  100
+ 177
```

13.
```
  340
   23
+   5
```

14.
```
   64
  520
+  13
```

15.
```
  200
  120
+  47
```

16.
```
    3
  220
  100
+ 146
  469
```

17.
```
  201
  383
   10
+ 301
```

18.
```
  212
  311
  203
+ 162
```

19.
```
  320
  200
  211
+   7
```

20.
```
   60
  201
  400
+  23
```

Adding Larger Numbers

To add numbers that are not lined up, first put the digits in columns.
Add only the digits with the same place values. Be sure to put your
answers in the correct columns.

Use These Steps

Add   1,006 + 41 + 2

1. Line up the digits in columns.

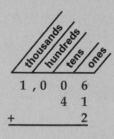

```
thousands hundreds tens ones
   1 , 0   0    6
           4    1
  +             2
```

2. Add the digits in each column, starting with the ones
place.

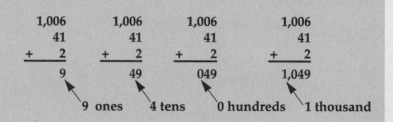

```
 1,006      1,006      1,006      1,006
    41         41         41         41
 +   2      +   2      +   2      +   2
 ___9       __49       _049       1,049
```

9 ones   4 tens   0 hundreds   1 thousand

Add.

1.
$$1,246 + 343 =$$
```
  1,246
+   343
  1,589
```

2.
$$3,300 + 500 =$$

3.
$$4,973 + 1,022 =$$

4.
$$6,811 + 25 + 100 =$$

5.
$$742 + 3 + 1,010 =$$

6.
$$2,422 + 163 + 4 =$$

7.
$$7,041 + 10 + 3 =$$

8.
$$3,006 + 2,021 + 340 =$$

9.
$$34 + 200 + 2,335 =$$

10.
$$5,143 + 736 + 100 =$$

11.
$$9,240 + 420 + 307 =$$

12.
$$2,040 + 712 + 5,133 =$$

Problem Solving: Using Rounding and Estimating

Many people and businesses make a monthly budget. They add up their expenses for one or more months. Then they estimate how much money they will spend in the coming months. A budget keeps them from spending more money than they make.

You do not need exact numbers to prepare a budget. You can use rounded numbers to add the numbers faster.

Example At the end of last month, Terry wrote down his expenses. He spent $63, $75, $39, and $51 on food. About how much should he plan to spend on food next month?

Step 1. Round each amount to the nearest ten.

$63 rounds to $60
$75 rounds to $80
$39 rounds to $40
$51 rounds to $50

Step 2. Add the rounded amounts.

$ 60
$ 80
$ 40
$ 50
─────
$230

Terry should plan to spend about $230 on food.

Week	Food	Clothes	Bus Fare	Other
1	$63	0	$5	$13
2	$75	$13	$7	$22
3	$39	$22	$12	$37
4	$51	$49	$19	$9

Round each amount to the nearest ten. Then add.

1. Terry spent $13, $22, and $49 for clothes. About how much should he plan to spend for clothes next month?

2. He spent $5, $7, $12, and $19 on bus fare. About how much should he plan to spend for bus fare next month?

Answer _____

Answer _____

3. Terry spent $13, $22, $37, and $9 on other expenses. About how much should he plan to spend for these expenses next month?

Answer _____

4. Does Terry spend more money on clothes or on bus fare?

Answer _____

5. Does Terry spend more money on food or on his other expenses?

Answer _____

6. Rent and utilities cost Terry about $500 per month. About how much did he spend all together last month on rent and utilities, food, and bus fare?

Answer _____

7. Terry's income is about $1,000 each month. He will be getting a $75 raise next month. About how much will his income be next month?

Answer _____

8. Terry saves about $50 each month. He can save $18 more each month when he gets his raise. About how much can he save each month after his raise?

Answer _____

9. Terry wants to buy a stereo with his savings. The stereo he wants to buy costs $239, plus $18 tax. About how much all together does the stereo cost?

Answer _____

10. Terry has saved $250. Does he have enough money to buy the stereo?

Answer _____

Mixed Review

Write the value of the underlined digit in each number.

1. 3<u>5</u> _____
2. <u>7</u>8 _____
3. 9<u>9</u> _____

4. 3<u>5</u>6 _____
5. <u>7</u>50 _____
6. <u>9</u>00 _____

7. 1,<u>2</u>47 _____
8. 5,<u>7</u>75 _____
9. <u>9</u>,043 _____

Round each number to the nearest ten.

10. 56 _____
11. 44 _____
12. 75 _____

13. 145 _____
14. 978 _____
15. 817 _____

16. 2,467 _____
17. 5,306 _____
18. 4,985 _____

Add.

19.
```
  23
+ 45
```

20.
```
  31
+  3
```

21.
```
  30
+ 48
```

22.
```
  65
   4
+ 20
```

23.
```
  33
  56
+ 10
```

24.
```
  48
  50
+  1
```

25. $46 + 20 =$

26. $52 + 22 =$

27. $30 + 39 + 10 =$

28. $54 + 31 + 12 =$

29.
```
  4,641
    223
+    11
```

30.
```
  3,010
    632
+ 1,225
```

31.
```
  1,025
    133
  7,520
+ 1,300
```

32.
```
    220
  3,210
  4,200
+ 2,259
```

33. $4,752 + 36 + 10 =$

34. $7,930 + 2,012 + 21 =$

35. $4 + 250 + 9,325 =$

Adding Two-Digit Numbers with Renaming

When you add digits in the ones column, the sum can sometimes be 10 or more. When this happens, rename by carrying to the tens column or the place to the left.

Use These Steps

Add   15 + 39

1. Line up the digits in columns.

```
 tens ones
  1    5
+ 3    9
```

2. Add the ones. 5 + 9 = 14 ones. Rename 14 ones as 1 ten and 4 ones. Write 4 ones in the ones column. Carry the 1 ten to the tens column.

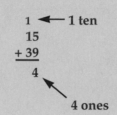

```
  1  ← 1 ten
 15
+39
  4
```
4 ones

3. Add the tens, plus the carried 1. 1 + 1 + 3 = 5 tens. Write 5 tens in the tens column.

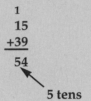

```
  1
 15
+39
 54
```
5 tens

Add.

1.
39 + 3 =
```
  1
 39
+ 3
 42
```

2.
27 + 27 =

3.
65 + 25 =

4.
87 + 3 =

5.
```
  47
+ 33
```

6.
```
  56
+  9
```

7.
```
  62
+ 18
```

8.
```
  19
+ 73
```

9.
```
  25
+  5
```

10.
```
  33
  26
+ 11
```

11.
```
  49
  28
+ 20
```

12.
```
  35
  21
+ 19
```

13.
```
  14
  38
+ 40
```

14.
```
  54
  18
+ 23
```

15.
36 + 48 + 1 =

16.
8 + 4 + 32 =

17.
6 + 13 + 64 =

18.
6 + 12 + 15 =

19.
3 + 18 + 22 =

20.
45 + 18 + 13 =

Adding Two-Digit Numbers with Renaming

When you add digits in the tens column, the sum can sometimes be 10 or more. When this happens, rename by carrying to the hundreds column or the place to the left.

Use These Steps

Add   96 + 38

1. Line up the digits in columns. Add the ones. 6 + 8 = 14 ones. Rename. Write 4 in the ones column. Carry the 1 ten to the tens column.

```
1  ← 1 ten
96
+38
 4
```
↙ 4 ones

2. Add the tens. 1 + 9 + 3 = 13 tens. Write 3 in the tens column.

```
 1
96
+38
34
```
↙ 3 tens

3. Write 1 in the hundreds column.

```
  1
 96
+38
134
```
↙ 1 hundred

Add.

1.
```
  1
 68
+47
115
```

2.
```
 74
+29
```

3.
```
 81
+39
```

4.
```
 93
+78
```

5.
```
 48
+96
```

6.   49 + 94 =

7.   38 + 91 =

8.   54 + 88 =

9.   55 + 97 =

10.   77 + 89 =

11.
```
 17
 99
+58
```

12.
```
 43
 19
+62
```

13.
```
 28
 93
+79
```

14.
```
 84
 47
+67
```

15.
```
 16
 98
+43
```

16.   45 + 23 + 55 =

17.   77 + 56 + 6 =

18.   76 + 1 + 19 =

Adding Three-Digit Numbers with Renaming

When you add three-digit numbers, you may need to rename two or more times.

Use These Steps

Add   372 + 499

1. Line up the digits in columns. Add the ones. 2 + 9 = 11 ones. Write 1 in the ones column. Carry 1 ten.

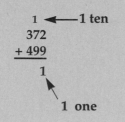

1 one

2. Add the tens. 1 + 7 + 9 = 17 tens. Write 7 in the tens column. Carry 1 hundred.

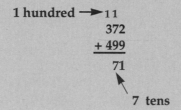

7 tens

3. Add the hundreds. 1 + 3 + 4 = 8 hundreds. Write 8 in the hundreds column.

```
 1 1
 372
+499
 871
```
8 hundreds

Add.

1.
```
 11
 349
+552
 901
```

2.
```
 878
+ 25
```

3.
```
 563
+449
```

4.
```
 981
+ 39
```

5.
```
 613
+387
```

6.   39 + 877 =

7.   886 + 29 =

8.   640 + 466 =

9.   72 + 379 =

10.
```
 1 1
 433
 297
+462
1,192
```

11.
```
 331
 828
+433
```

12.
```
 849
 908
+884
```

13.
```
 119
 488
+281
```

14.
```
 670
 821
+943
```

15.   561 + 849 + 999 =

16.   189 + 248 + 577 =

17.   251 + 629 + 125 =

18.   159 + 623 + 851 =

43

Adding Larger Numbers with Renaming

When you add the thousands column, you may need to rename by carrying to the next column or place to the left.

Add   1,872 + 128

1. Line up the digits in columns. Add the ones. $2 + 8 = 10$ ones. Write 0 in the ones column. Carry 1 ten.

$$\begin{array}{r} 1 \leftarrow 1 \text{ ten} \\ 1,872 \\ +\ 128 \\ \hline 0 \end{array}$$

0 ones

2. Add the tens. $1 + 7 + 2 = 10$ tens. Write 0 in the tens column. Carry 1 hundred.

$$\begin{array}{r} 1 \text{ hundred} \rightarrow 11 \\ 1,872 \\ +\ 128 \\ \hline 00 \end{array}$$

0 tens

3. Add the hundreds. $1 + 8 + 1 = 10$ hundreds. Write 0 in the hundreds column. Write 1 in the thousands column. Add the thousands. Write 2 in the thousands column.

$$\begin{array}{r} 1 \text{ thousand} \rightarrow 1\,11 \\ 1,872 \\ +\ 128 \\ \hline 2,000 \end{array}$$

0 hundreds
2 thousands

Add.

1.
$$\begin{array}{r} 1\ 11 \\ 1,938 \\ +\ \ 289 \\ \hline 2,227 \end{array}$$

2.
$$\begin{array}{r} 2,579 \\ +\ \ 164 \\ \hline \end{array}$$

3.
$$\begin{array}{r} 769 \\ +\ 433 \\ \hline \end{array}$$

4.
$$\begin{array}{r} 821 \\ +\ 399 \\ \hline \end{array}$$

5.
$$\begin{array}{r} 3,352 \\ +\ \ 190 \\ \hline \end{array}$$

6. $4,882 + 595 =$

7. $1,379 + 845 =$

8. $8,950 + 959 =$

9. $8,987 + 7,134 =$

10.
$$\begin{array}{r} 6,826 \\ 3,891 \\ +\ 8,717 \\ \hline \end{array}$$

11.
$$\begin{array}{r} 8,686 \\ 2,982 \\ +\ 6,531 \\ \hline \end{array}$$

12.
$$\begin{array}{r} 9,997 \\ 3,987 \\ +\ 4,003 \\ \hline \end{array}$$

13.
$$\begin{array}{r} 14,970 \\ 8,206 \\ +\ 7,334 \\ \hline \end{array}$$

14.
$$\begin{array}{r} 19,986 \\ 11,861 \\ +\ 19,374 \\ \hline \end{array}$$

15. $3,567 + 14,556 + 435 =$

16. $45,679 + 23,440 + 2,400 =$

Adding Zeros

Zero plus any number equals that number. Remember to add
carried numbers.

Add.

1.
```
    1
  304
+ 106
─────
  410
```

2.
```
  409
+ 301
─────
```

3.
```
 1,308
+    3
──────
```

4.
```
   907
+ 1,005
───────
```

5.
```
  5,601
+ 1,109
───────
```

6. $5,096 + 68 =$

7. $6,004 + 209 =$

8. $908 + 84 =$

9. $6,808 + 204 =$

10.
```
  9,006
    804
+     7
───────
```

11.
```
  17,007
   1,009
+    102
────────
```

12.
```
  14,908
   9,009
+  3,000
────────
```

13.
```
  9,001
    409
+   800
───────
```

14.
```
  8,601
  4,009
+ 2,000
───────
```

15. $22,005 + 409 + 1,000 =$

16. $45,000 + 3,005 + 509 =$

Adding Long Columns

When you add columns of four or more numbers, it helps to add the numbers in steps.

Use These Steps

Add   16 + 18 + 82 + 49 + 61

1. Line up the digits in columns. Add groups of digits in the ones column.

```
   2
  16 ────── 6 + 8 = 14
  18
  82 ──┐ 2 + 9 + 1 = 12
  49   ┘          26
 +61
   6
```

2. Add groups of digits in the tens column.

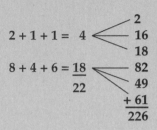

```
                        2
2 + 1 + 1 =  4 ◄──── 16
                       18
8 + 4 + 6 = 18 ◄──── 82
            22         49
                     + 61
                      226
```

Add.

1.
```
    6
    4
    3
    9
 +  7
   29
```

2.
```
    7
    8
    5
    5
 +  2
```

3.
```
   41
   19
   63
   27
 + 90
```

4.
```
   88
   19
   13
   41
 + 62
```

5.
```
   430
   762
   148
   299
 + 107
```

6.
```
   827
    33
   143
    69
 +  20
```

7.
```
  1,285
     95
    140
      7
 + 6,581
```

8.
```
  4,281
  5,479
  3,207
  1,800
 + 1,021
```

9.
```
  2,455
  4,532
  3,658
  3,275
 + 8,103
```

10.
```
  3,559
    411
  6,149
  3,622
 +  981
```

11.
```
    675
     25
  4,122
    318
 +  245
```

12.
```
   3,454
  12,357
  44,113
   2,395
 + 7,001
```

Place Value to 1,000,000

The chart on the right shows place value to the millions place.

You read the number 1,476,901 as one million, four hundred seventy-six thousand, nine hundred one.

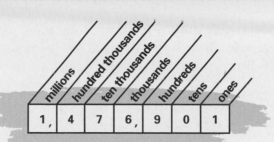

Use These Steps

Write the place values of each digit in the number 42,530.

1. Write the digits and their place values.

4 ten thousands	
2 thousands	
5 hundreds	
3 tens	
0 ones	

2. Write the values of the digits.

40,000
2,000
500
30
0

Write the place value for each number. Then write the value of each digit.

1. 3,472,400

3	millions	3,000,000
4	hundred thousands	400,000
7	ten thousands	70,000
2	thousands	2,000
4	hundreds	400
0	tens	00
0	ones	0

2. 6,480,603

6		
4		
8		
0		
6		
0		
3		

Write the value of the underlined digit in each number.

3. 3<u>4</u>5 _____40_____

4. 10<u>3</u> _____

5. <u>4</u>,578 _____

6. 2,<u>9</u>90 _____

7. 2<u>4</u>,704 _____

8. 56,<u>9</u>04 _____

9. <u>2</u>34,004 _____

10. 79<u>3</u>,456 _____

11. <u>3</u>,579,844 _____

Adding Larger Numbers

Adding larger numbers is the same as adding small numbers. Be sure that the digits with the same place values are lined up, and that you have not forgotten any digits.

$$4,749,200 + 7,761,035$$

$$\begin{array}{r} 4,749,200 \\ + 7,761,035 \\ \hline 12,510,235 \end{array}$$

Use These Steps

Add 998,206 + 3,044,052

1. Line up the digits in columns.

$$\begin{array}{r} 998,206 \\ + 3,044,052 \end{array}$$

2. Add, beginning with the digits in the ones place.

$$\begin{array}{r} {\scriptstyle 1\ 11} \\ 998,206 \\ + 3,044,052 \\ \hline 4,042,258 \end{array}$$

Add.

1.
$$\begin{array}{r} {\scriptstyle 1\ 1} \\ 20,473 \\ + 99,068 \\ \hline 119,541 \end{array}$$

2.
$$\begin{array}{r} 38,468 \\ + 106,810 \end{array}$$

3.
$$\begin{array}{r} 68,024 \\ + 89,118 \end{array}$$

4.
$$\begin{array}{r} 1,061,247 \\ + \ \ \ 761,785 \end{array}$$

5. $1,460,000 + 39,000 =$

6. $793,047 + 3,799 =$

7. $62,901 + 439 =$

8.
$$\begin{array}{r} 93,486 \\ 129,882 \\ + 1,937,246 \end{array}$$

9.
$$\begin{array}{r} 98,621 \\ 261,084 \\ + 2,689,677 \end{array}$$

10.
$$\begin{array}{r} 237,069 \\ 61,488 \\ + 1,968,984 \end{array}$$

11.
$$\begin{array}{r} 62,949 \\ 981,098 \\ + 6,789,987 \end{array}$$

12. $372,000 + 1,000,497 =$

13. $45,446 + 12,530 + 3,445 =$

14. $3,998 + 133,404 + 12,001 =$

15. $45,978 + 3,166 + 298,566 =$

Problem Solving: Using Rounding and Estimating

When we estimate, we usually round numbers to their greatest place value. The greatest place value is the place value of the first digit on the left, called the *lead digit*.

> 1,575 rounds to 2,000
> 925 rounds to 900
> 39 rounds to 40

Example   The Kelleys drove from Los Angeles to Santa Fe. They kept a record of how many miles they drove, how many gallons of gas they used, and how much money they spent.

Look at the chart. About how many miles did the Kelleys drive the first 2 days of their trip?

Day	Miles	Gallons of Gas	Expenses
1	97	11	$56
2	123	13	$80
3	180	16	$94
4	193	19	$96
5	245	22	$59

▶ **Step 1.**   Round the miles driven each day. To round, look at the digit next to the lead digit.

97   123

If the digit to the right of the lead digit is 5 or greater, add 1 to the lead digit and change the other digits to 0. If the digit to the right of the lead digit is less than 5, change all digits except the lead digit to 0.

97 rounds to 100
123 rounds to 100

▶ **Step 2.**   Add the rounded numbers.

$$\begin{array}{r} 100 \\ + 100 \\ \hline 200 \end{array}$$

The Kelleys drove about 200 miles the first 2 days of their trip.

Solve by rounding to the lead digit.

1. About how many miles did the Kelleys drive on the last 3 days of their trip?

   Answer _____

2. About how many miles in all did they travel from Los Angeles to Santa Fe?

   Answer _____

3. About how many gallons of gas did they use on the trip?

   Answer _____

4. Estimate how much money the Kelleys spent on the trip.

   Answer _____

5. The Kelleys estimated they would spend $500 on their trip. Did they spend more or less than they estimated?

   Answer _____

6. Last year the Kelleys drove 1,927 miles on a trip. Round this number to the lead digit.

   Answer _____

7. Which year did the Kelleys drive farther: last year or this year?

   Answer _____

8. Next year the Kelleys plan to drive 2,794 miles from Los Angeles to New York City. Round this number to the lead digit.

   Answer _____

9. About how many total miles will the round trip be from Los Angeles to New York City and back?

   Answer _____

10. Which year will the Kelleys drive farther: this year or next year?

    Answer _____

Unit 2 *Review*

Add.

1.
$$\begin{array}{r} 6 \\ +\,5 \\ \hline \end{array}$$

2.
$$\begin{array}{r} 8 \\ +\,2 \\ \hline \end{array}$$

3.
$$\begin{array}{r} 7 \\ +\,4 \\ \hline \end{array}$$

4.
$$\begin{array}{r} 7 \\ 2 \\ +\,3 \\ \hline \end{array}$$

5.
$$\begin{array}{r} 8 \\ 9 \\ +\,1 \\ \hline \end{array}$$

6.
$$\begin{array}{r} 5 \\ 3 \\ +\,2 \\ \hline \end{array}$$

7.
$$\begin{array}{r} 89 \\ +\,10 \\ \hline \end{array}$$

8.
$$\begin{array}{r} 72 \\ +\,23 \\ \hline \end{array}$$

9.
$$\begin{array}{r} 53 \\ +\,42 \\ \hline \end{array}$$

10.
$$\begin{array}{r} 69 \\ 20 \\ +\,10 \\ \hline \end{array}$$

11.
$$\begin{array}{r} 31 \\ 17 \\ +\,41 \\ \hline \end{array}$$

12.
$$\begin{array}{r} 35 \\ 10 \\ +\,54 \\ \hline \end{array}$$

13. $40 + 24 =$

14. $76 + 21 =$

15. $30 + 11 + 27 =$

16. $50 + 29 + 10 =$

17.
$$\begin{array}{r} 309 \\ +\ \ 40 \\ \hline \end{array}$$

18.
$$\begin{array}{r} 764 \\ +\,135 \\ \hline \end{array}$$

19.
$$\begin{array}{r} 41 \\ 7 \\ +\,401 \\ \hline \end{array}$$

20.
$$\begin{array}{r} 34 \\ 253 \\ +\,1{,}710 \\ \hline \end{array}$$

21.
$$\begin{array}{r} 45 \\ 843 \\ +\,6{,}010 \\ \hline \end{array}$$

22. $4{,}037 + 2{,}260 =$

23. $5{,}444 + 434 + 21 =$

24. $100 + 2{,}405 + 50 =$

25.
$$\begin{array}{r} 27 \\ +\,63 \\ \hline \end{array}$$

26.
$$\begin{array}{r} 56 \\ +\,39 \\ \hline \end{array}$$

27.
$$\begin{array}{r} 52 \\ +\,88 \\ \hline \end{array}$$

28.
$$\begin{array}{r} 19 \\ 73 \\ +\ \ 3 \\ \hline \end{array}$$

29.
$$\begin{array}{r} 25 \\ 75 \\ +\ \ 5 \\ \hline \end{array}$$

30.
$$\begin{array}{r} 57 \\ 19 \\ +\,44 \\ \hline \end{array}$$

31. $39 + 68 =$

32. $27 + 27 =$

33. $65 + 75 + 5 =$

34. $47 + 23 + 93 =$

35.
```
  268
+  47
```

36.
```
9,974
+   29
```

37.
```
 1,381
+ 8,739
```

38.
```
  693
  378
+  29
```

39.
```
  948
   96
+ 340
```

40. 849 + 94 =

41. 3,938 + 91 =

42. 1,354 + 788 + 21 =

43.
```
9,006
  804
+   7
```

44.
```
17,007
 1,009
+  102
```

45.
```
14,908
 9,009
+ 3,000
```

46.
```
  430
  762
  148
  299
+ 107
```

47.
```
  827
   33
  143
   69
+  20
```

48.
```
 56,978
+ 49,078
```

49.
```
 345,750
+ 742,283
```

50.
```
 3,740,033
+ 4,005,778
```

51.
```
 8,798,023
+ 3,375,611
```

52.
```
 1,569,756
+   865,599
```

53. 53,890 + 4,877 =

54. 29,000 + 567,854 + 4,321 =

Below is a list of the problems in this review and the pages on which the skills are taught. If you missed any problems, turn to the pages listed and practice the skills. Then correct the problems you missed in the Unit Review.

Unit 3 — SUBTRACTING WHOLE NUMBERS

You subtract whole numbers when you take away one amount from another amount. You use subtraction to see how much money you have left after you pay bills, to find how many more days are left in the year, or to learn the difference between scores in a baseball game.

In this unit, you will learn how to set up a problem to subtract whole numbers. You will also learn how to borrow, how to subtract with zeros, and how to check answers by using addition.

Getting Ready

You should be familiar with the skills on this page and the next before you begin this unit. To check your answers, turn to page 182.

 When you are working with whole numbers, you need to know the place value of each digit.

The place value chart shows the value of each digit in the number ninety-three thousand, four hundred twenty-seven. Write the value of each digit.

1. 93,427

   9 ten thousands

   _____

   _____

   _____

   _____

2. 42,076

   _____

   _____

   _____

   _____

   _____

hundred thousands	ten thousands	thousands	hundreds	tens	ones
	9	3,	4	2	7

For review, see Unit 2, page 47.

To set up an addition problem, first line up the digits in columns. Then solve the problem.

Line up the following problems and solve.

3.
467 + 31 =

```
  467
+  31
-----
  498
```

4.
346 + 29 =

5.
4,960 + 321 =

6.
8,927 + 1,007 =

7.
1,682 + 851 =

8.
605 + 8,126 =

9.
720 + 809 =

10.
2,016 + 8 =

For review, see Unit 2, pages 43-45.

To round a number, you need to know the place value of each digit.

Round each number to the nearest ten.

11. 29   30   **12.** 155   _____   **13.** 406   _____   **14.** 1,299   _____

Round each number to the nearest hundred.

15. 430   _____   **16.** 792   _____   **17.** 1,007   _____   **18.** 5,383   _____

Round each number to the nearest thousand.

19. 5,830   _____   **20.** 9,436   _____   **21.** 15,099   _____   **22.** 99,650   _____

For review see, Unit 1, pages 20-22.

In addition, it doesn't matter which number is written first. In subtraction, the larger number must go on top.

Compare the following numbers. Circle the larger number.

23. 30   (300)   **24.** 28   18   **25.** 37   73

26. 297   279   **27.** 806   860   **28.** 400   4,000

For review, see Unit 1, page 13.

Subtraction Facts

To subtract larger numbers, you should first know the basic subtraction facts. You will find it helpful to know the following facts by heart.

Subtract the following numbers to complete each row.
Notice that the answers form a pattern.

1.

0	1	2	3	4	5	6	7	8	9
-0	-0	-0	-0	-0	-0	-0	-0	-0	-0
0	1	2	3	4	5	6	7	8	9

2.

1	2	3	4	5	6	7	8	9	10
-1	-1	-1	-1	-1	-1	-1	-1	-1	-1

3.

2	3	4	5	6	7	8	9	10	11
-2	-2	-2	-2	-2	-2	-2	-2	-2	-2

4.

3	4	5	6	7	8	9	10	11	12
-3	-3	-3	-3	-3	-3	-3	-3	-3	-3

5.

4	5	6	7	8	9	10	11	12	13
-4	-4	-4	-4	-4	-4	-4	-4	-4	-4

6.

5	6	7	8	9	10	11	12	13	14
-5	-5	-5	-5	-5	-5	-5	-5	-5	-5

7.

6	7	8	9	10	11	12	13	14	15
-6	-6	-6	-6	-6	-6	-6	-6	-6	-6

8.

7	8	9	10	11	12	13	14	15	16
-7	-7	-7	-7	-7	-7	-7	-7	-7	-7

9.

8	9	10	11	12	13	14	15	16	17
-8	-8	-8	-8	-8	-8	-8	-8	-8	-8

10.

9	10	11	12	13	14	15	16	17	18
-9	-9	-9	-9	-9	-9	-9	-9	-9	-9

Subtraction Facts Practice

You can use the addition facts table to complete subtraction facts, since addition and subtraction are opposite operations.

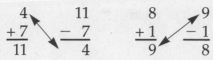

$$\begin{array}{c}4\\+7\\\hline 11\end{array} \qquad \begin{array}{c}11\\-7\\\hline 4\end{array} \qquad \begin{array}{c}8\\+1\\\hline 9\end{array} \qquad \begin{array}{c}9\\-1\\\hline 8\end{array}$$

Use These Steps

Complete the subtraction fact 6 − 4.

1. First, find the smaller number, 4, in the farthest column on the left.

2. Next, move to the right along that row until you find the larger number, 6.

3. Then move to the top of the column to find the answer, 2.

−	0	1	**2**	3	4	5	6	7	8	9
0	0	1	2	3	4	5	6	7	8	9
1	1	2	3	4	5	6	7	8	9	10
2	2	3	4	5	6	7	8	9	10	11
3	3	4	5	6	7	8	9	10	11	12
4	4	5	**6**	7	8	9	10	11	12	13
5	5	6	7	8	9	10	11	12	13	14
6	6	7	8	9	10	11	12	13	14	15
7	7	8	9	10	11	12	13	14	15	16
8	8	9	10	11	12	13	14	15	16	17
9	9	10	11	12	13	14	15	16	17	18

Use the table to complete the following subtraction facts.

1. $12 − 6 = 6$
2. $15 − 8 =$
3. $17 − 9 =$
4. $7 − 3 =$
5. $18 − 9 =$
6. $11 − 6 =$

7. $10 − 5 =$
8. $3 − 0 =$
9. $14 − 8 =$
10. $12 − 9 =$
11. $15 − 6 =$
12. $16 − 7 =$

13. $8 − 6 =$
14. $10 − 2 =$
15. $9 − 4 =$
16. $13 − 5 =$
17. $16 − 0 =$
18. $9 − 9 =$

Subtraction Facts Practice

Subtraction problems can be written vertically or horizontally.

minuend $\longrightarrow$ 9
subtrahend $\longrightarrow$ $\underline{-6}$ is the same as $9 - 6 = 3$
difference $\longrightarrow$ 3

Complete the following subtraction facts. You can use the table on page 56 if you need help remembering the facts.

1. $15 - 7 = \boxed{8}$

2. $8 - 4 = \square$

3. $10 - 6 = \square$

4. $9 - 3 = \square$

5. $7 - \boxed{4} = 3$

6. $12 - \square = 4$

7. $14 - \square = 7$

8. $6 - \square = 6$

9. $\boxed{11} - 2 = 9$

10. $\square - 6 = 5$

11. $\square - 9 = 1$

12. $\square - 8 = 2$

13. $6 - 2 = \square$

14. $9 - 4 = \square$

15. $\square - 3 = 2$

16. $12 - \square = 6$

17. $\square - 8 = 7$

18. $15 - 6 = \square$

19. $13 - \square = 6$

20. $\square - 4 = 3$

21. $5 - 0 = \square$

22. $7 - \square = 5$

23. $18 - \square = 9$

24. $17 - 9 = \square$

25. $10 - \square = 2$

26. $4 - \square = 3$

27. $14 - \square = 14$

28. $\square - 1 = 6$

29. $\square - 2 = 8$

30. $16 - \square = 7$

31. $9 - 9 = \square$

32. $\square - 4 = 1$

Fill in the boxes with any numbers that make the differences. There may be more than one set of numbers that makes a true statement.

33. $\boxed{9} - \boxed{2} = 7$

34. $\square - \square = 9$

35. $\square - \square = 5$

36. $\square - \square = 0$

37. $\square - \square = 4$

38. $\square - \square = 8$

39. $\square - \square = 1$

40. $\square - \square = 6$

41. $\square - \square = 2$

Subtraction as the Opposite of Addition

Subtracting whole numbers is the opposite of adding whole numbers. This means that you can check the answer to a subtraction problem by adding the answer to the number you subtracted. The sum should be the same as the top number.

$$
\begin{array}{r} 6 \\ -4 \\ \hline 2 \end{array}
\qquad
\begin{array}{r} 2 \\ +4 \\ \hline 6 \end{array}
$$

Use These Steps

Subtract $17 - 8$

1. Line up the digits.

$$\begin{array}{r} 17 \\ -\ 8 \\ \hline \end{array}$$

2. Subtract.

$$\begin{array}{r} 17 \\ -\ 8 \\ \hline 9 \end{array}$$

3. Check by adding the answer, 9, to the bottom number, 8. The sum should be the same as the top number, 17.

Check:

$$
\begin{array}{r} 17 \\ -\ 8 \\ \hline 9 \end{array}
\qquad
\begin{array}{r} 9 \\ +8 \\ \hline 17 \end{array}
$$

Subtract. Use addition to check your answers.

1.
$$
\begin{array}{r} 12 \\ -\ 3 \\ \hline 9 \end{array}
\quad
\begin{array}{r} 9 \\ +3 \\ \hline 12 \end{array}
$$

2.
$$\begin{array}{r} 8 \\ -7 \\ \hline \end{array}$$

3.
$$\begin{array}{r} 6 \\ -0 \\ \hline \end{array}$$

4.
$$\begin{array}{r} 10 \\ -\ 2 \\ \hline \end{array}$$

5.
$$\begin{array}{r} 13 \\ -\ 5 \\ \hline \end{array}$$

6.
$$\begin{array}{r} 10 \\ -\ 6 \\ \hline \end{array}$$

7.
$$\begin{array}{r} 9 \\ -0 \\ \hline \end{array}$$

8.
$$\begin{array}{r} 16 \\ -\ 8 \\ \hline \end{array}$$

9.
$$\begin{array}{r} 14 \\ -\ 5 \\ \hline \end{array}$$

10.
$$\begin{array}{r} 9 \\ -6 \\ \hline \end{array}$$

11. $15 - 7 =$

12. $2 - 0 =$

13. $12 - 7 =$

14. $16 - 9 =$

15. $16 - 5 =$

16. $11 - 4 =$

17. $18 - 9 =$

18. $13 - 7 =$

Real-Life Application   Daily Living

The chart below shows the high and low temperatures on one day in September for twelve cities in the United States.

Example What was the difference between the low and high temperatures for Albany?

$$\begin{array}{r} 63 \\ -\,43 \\ \hline 20 \end{array}$$

The difference was 20 degrees.

Temperature in Degrees Fahrenheit		
City	High	Low
Albany	63	43
Atlanta	71	54
Boston	62	51
Chicago	60	36
Denver	81	43
Honolulu	89	72
Las Vegas	97	72
Miami	88	81
Omaha	69	42
St. Louis	63	41
Spokane	77	49
Wichita	64	44

Solve. Be sure to put the greater number on top when you subtract.

1. Find the difference between the low and high temperatures for Boston.

   Answer _____

2. What was the difference between the high and low temperatures for Omaha?

   Answer _____

3. How much higher was the high temperature for Las Vegas than the high temperature for Chicago?

   Answer _____

4. How much lower was the low temperature for Chicago than the low temperature for Spokane?

   Answer _____

5. How much higher was the high temperature for Miami than the low temperature for Wichita?

   Answer _____

6. How much lower was the low temperature for St. Louis than the high temperature for Denver?

   Answer _____

Subtracting from Two-Digit Numbers

Use the subtraction facts to subtract two-digit numbers. Be sure to line up your answers in the correct columns. You may want to use the table on page 56 to help you.

Use These Steps

Subtract 57 − 36

1. Be sure that the ones and the tens are lined up.

2. Subtract the ones. 7 − 6 = 1 one. Put the 1 one in the ones column.

3. Subtract the tens. 5 − 3 = 2 tens. Put the 2 tens in the tens column. Check your answer by adding.

```
     57
  −  36
```

```
     57
  −  36
      1
```
↖ 1 one

```
     57          Check:
  −  36            21
     21          + 36
                   57
```
↖ 2 tens

Subtract. Use addition to check your answers.

1.
```
    78       44
  − 34     + 34
    44       78
```

2.
```
    89
  − 72
```

3.
```
    42
  − 21
```

4.
```
    96
  − 43
```

5.
```
    68
  − 25
```

6.
```
    49
  − 37
```

7.
```
    32
  − 11
```

8.
```
    99
  − 54
```

9.
```
    58
  − 30
```

10.
```
    22
  − 10
```

11.
```
    87
  − 20
```

12.
```
    51
  − 30
```

13.
```
    86
  − 36
    50
```

14.
```
    79
  − 39
```

15.
```
    60
  − 50
```

16.
```
    28
  − 23
```

17.
```
    48
  − 17
```

18.
```
    54
  − 22
```

19.
```
    63
  − 40
```

20.
```
    72
  − 32
```

Subtracting from Two-Digit Numbers

To subtract numbers that are not lined up, first put the digits in columns. Line up the digits that have the same place value.

Use These Steps

Subtract 25 − 5

1. Write the digits in columns so that the digits with the same place value are lined up.

$$\begin{array}{r} 25 \\ -\ 5 \\ \hline \end{array}$$

2. Subtract the ones.
5 − 5 = 0 ones.

$$\begin{array}{r} 25 \\ -\ 5 \\ \hline 0 \end{array}$$

↖ 0 ones

3. Subtract the tens. There are no (0) tens in the bottom number.
2 − 0 = 2 tens. Check.

Check:

$$\begin{array}{r} 25 \\ -\ 5 \\ \hline 20 \end{array} \qquad \begin{array}{r} 20 \\ +\ 5 \\ \hline 25 \end{array}$$

↖ 2 tens

Subtract. Use addition to check your answers.

1.
36 − 5 =

$$\begin{array}{r} 36 \\ -\ 5 \\ \hline 31 \end{array} \qquad \begin{array}{r} 31 \\ +\ 5 \\ \hline 36 \end{array}$$

2.
77 − 4 =

3.
99 − 6 =

4.
83 − 2 =

5.
18 − 6 =

6.
29 − 7 =

7.
58 − 4 =

8.
99 − 5 =

9. Simon's Pet Store had a shipment of 29 goldfish on Monday. By Saturday, 8 of the goldfish had been bought. How many goldfish were left?

10. Simon's got 46 guppies in the same shipment. By Saturday, 4 of them had been sold. How many guppies were left?

Answer_____

Answer_____

Subtracting from Larger Numbers

To subtract larger numbers, first be sure that the digits with the same place value are lined up. Subtract only the digits with the same place value. Be sure to put your answers in the correct columns.

Use These Steps

Subtract
$$\begin{array}{r} 459 \\ -\ 341 \end{array}$$

1. Be sure that the digits are lined up.

$$\begin{array}{r} 459 \\ -\ 341 \end{array}$$

2. Subtract each column, starting with the digits in the ones place.

$$\begin{array}{r} 459 \\ -\ 341 \\ \hline 8 \end{array}$$
8 ones

$$\begin{array}{r} 459 \\ -\ 341 \\ \hline 18 \end{array}$$
1 ten

$$\begin{array}{r} 459 \\ -\ 341 \\ \hline 118 \end{array}$$
1 hundred

Check:
$$\begin{array}{r} 118 \\ +\ 341 \\ \hline 459 \end{array}$$

Subtract. Use addition to check your answers.

1.
$$\begin{array}{r} 920 \\ -\ 500 \\ \hline 420 \end{array} \qquad \begin{array}{r} 420 \\ +\ 500 \\ \hline 920 \end{array}$$

2.
$$\begin{array}{r} 876 \\ -\ 761 \end{array}$$

3.
$$\begin{array}{r} 2,398 \\ -\ 1,233 \end{array}$$

4.
$$\begin{array}{r} 1,444 \\ -\ 1,243 \end{array}$$

5.
$$\begin{array}{r} 123 \\ -\ 23 \end{array}$$

6.
$$\begin{array}{r} 436 \\ -\ 1 \end{array}$$

7.
$$\begin{array}{r} 7,980 \\ -\ 920 \end{array}$$

8.
$$\begin{array}{r} 5,874 \\ -\ 3 \end{array}$$

9. $3,253 - 1,132 =$
$$\begin{array}{r} 3,253 \\ -\ 1,132 \\ \hline 2,121 \end{array} \qquad \begin{array}{r} 2,121 \\ +\ 1,132 \\ \hline 3,253 \end{array}$$

10. $4,899 - 236 =$

11. $2,789 - 77 =$

12. By noon yesterday, the Waverly highway crew had finished paving 1,400 feet of roadway. The entire roadway is 1,456 feet long. How many feet do they have left to pave?

13. The West Branch Library made $465 at its book sale on Saturday and $786 on Sunday. How much more did the library make on Sunday than on Saturday?

Answer_____

Answer_____

Mixed Review

Add or subtract.

1.
$$2 + 1$$

2.
$$8 - 0$$

3.
$$15 - 7$$

4.
$$8 + 6$$

5.
$$8 + 9$$

6.
$$10 - 3$$

7.
$$7 + 4$$

8. $16 - 9 =$

9. $7 + 8 =$

10. $10 - 1 =$

11. $9 + 4 =$

12.
$$71 + 28$$

13.
$$89 - 86$$

14.
$$54 + 40$$

15.
$$43 - 23$$

16. $36 + 13 =$

17. $77 - 54 =$

18. $50 + 10 =$

19. $64 - 30 =$

20.
$$123 + 23$$

21.
$$434 - 1$$

22.
$$7{,}980 + 1{,}019$$

23.
$$1{,}987 - 3$$

24. $6{,}542 - 2{,}111 =$

25. $2{,}699 + 200 =$

26. $5{,}196 - 71 =$

27. Sam bought a jacket for $15 and a pair of pants for $15 at the school rummage sale. How much did he spend in all?

28. Sam paid for the clothes with $40. How much change did he get back?

Answer_____

Answer_____

63

Problem Solving: Using a Step-by-Step Plan

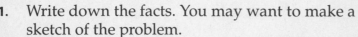

Solving word problems is easier if you follow a step-by-step plan. First, read the problem several times until you understand what the problem is about. Then follow these steps.

1. Write down the facts. You may want to make a sketch of the problem.
2. Think about what you need to do. Decide if you need to add or subtract, or both.
3. Set up a number problem.
4. Solve.
5. Write the answer to the question.

Example This week Marty is building a fence for his back yard. The yard is in the shape of a rectangle. Two sides of the yard are each 40 feet long. The other sides are each 27 feet long. How many feet of fencing material will Marty need?

▶ **Step 1.** Write down the facts.

40 feet long on two sides
27 feet long on two sides

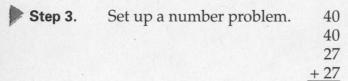

▶ **Step 2.** Decide what to do.

Since you need a total length, you will need to add the lengths of the four sides of the yard.

▶ **Step 3.** Set up a number problem.

$$
\begin{array}{r}
40 \\
40 \\
27 \\
+\,27 \\
\end{array}
$$

▶ **Step 4.** Solve.

$$
\begin{array}{r}
40 \\
40 \\
27 \\
+\,27 \\
\hline
134 \\
\end{array}
$$

▶ **Step 5.** Answer the question.

Marty will need 134 feet of fencing material.

Use the steps on page 64 to solve each problem.

1. By installing storm windows in her house, Carolyn can save $1,850 dollars on her heating bill over the next ten years. Marty is installing these windows for her at a cost of $730. After paying for Marty's work, how much money will Carolyn actually save?

   Step 1.

   Step 2.

   Step 3.

   Step 4.

   Step 5.

2. Della Edwards raises Siamese cats to sell to pet stores. In 1989 she sold 36 cats. In 1990 she sold 52 cats. She sold 72 cats in 1991, and 41 cats in 1992. How many cats all together did Della sell in four years?

   Step 1.

   Step 2.

   Step 3.

   Step 4.

   Step 5.

Subtracting from Two-Digit Numbers with Renaming

When you are subtracting, the digit you are subtracting from can sometimes be too small. When this happens, borrow 1 ten from the next column to the left. Rename 1 ten as 10 ones.

Use These Steps

Subtract 42
          − 19

1. Since you can't subtract 9 from 2, borrow 1 ten. Cross out the 4 and write a 3 above it.

2. Rename the borrowed ten as ten ones. 2 + 10 = 12. Cross out the 2 and write a 12 above it.

3. Subtract the ones. 12 − 9 = 3 ones. Subtract the tens. 3 − 1 = 2 tens. Check your answer.

```
      3                    3 12              3 12        Check:
      4 2                  4 2               4 2              1
    − 1 9                − 1 9             − 1 9           2 3
                                            2 3          + 1 9
                                                          4 2
```

Subtract. Use addition to check your answers.

```
1.   3 16        1          2.           3.           4.
     4 6         1 8            92           67           71
   − 2 8       + 2 8         − 7 3        − 4 9        − 3 3
     1 8         4 6
```

```
5.               6.           7.           8.
     53              32           23           84
   −  8           −  7         −  6         −  9
```

```
9.               10.          11.          12.
     45               62           24           96
   − 37             − 59         − 15         − 88
```

```
13.              14.          15.          16.
     31               77           56           23
   −  6             − 49         −  8         − 17
```

```
17.              18.          19.          20.
     45               62           74           86
   −  6             − 55         −  9         − 18
```

66

Subtracting from Two-Digit Numbers with Renaming

Line up the numbers carefully. The larger number goes on top. Borrow from the tens column only if you can't subtract the digit in the ones column.

Use These Steps

Subtract 86 − 78

1. **Line up the digits in columns.**

2. **Since you can't subtract 8 from 6, borrow 1 ten. Rename.**

3. **Subtract the ones. 16 − 8 = 8 ones. Subtract the tens. 7 − 7 = 0 tens. Check.**

			Check:
	7 16	7 16	1
86	8̸ 6̸	8̸ 6̸	8
− 78	− 7 8	− 7 8	+ 7 8
		8	8 6

Subtract. Use addition to check your answers.

1.
```
  3 14
  4̸ 4̸        18
 − 26       + 26
   18         44
```

2.
```
  53
 − 34
```

3.
```
  94
 − 79
```

4.
```
  82
 − 48
```

5. 52 − 43 =

6. 72 − 59 =

7. 34 − 26 =

8. 91 − 87 =

9. 82 − 9 =

10. 66 − 7 =

11. 47 − 8 =

12. 38 − 9 =

13. A tune-up at Mike's Repair Shop costs $45. A tune-up kit and spark plugs to do the tune-up yourself cost $28. How much will you save if you do the work yourself?

14. The A & A Auto Shop charges a total of $74 to replace a car muffler. If the muffler alone costs $46, how much can you save if you do the work yourself?

Answer _____

Answer _____

Zeros in Subtraction

To subtract from zero in the ones place, rename just as you have done before.

Use These Steps

Subtract   20
           − 13

1. Be sure that the digits are lined up.

2. Since you can't subtract 3 from 0, borrow 1 ten. Rename. Cross out the 2 and write a 1 above it. Cross out the 0 and write a 10 above it.

3. Subtract the ones. 10 − 3 = 7 ones. Subtract the tens. 1 − 1 = 0 tens. Check your answer.

                                              Check:

```
                        1 10           1 10        1
   20                    2 0            2 0         7
 − 13                  − 1 3          − 1 3       + 1 3
                                          7         2 0
```

Subtract. Use addition to check your answers.

1.
```
 2 10
  3 0        17
− 13       + 13
  17         30
```

2.
```
  50
− 26
```

3.
```
  70
− 39
```

4.
```
  40
− 14
```

5.
```
  60
− 54
```

6.
```
  80
− 73
```

7.
```
  90
− 88
```

8.
```
  20
− 11
```

9.
```
  30
−  4
```

10.
```
  50
−  7
```

11.
```
  70
−  2
```

12.
```
  90
−  9
```

13.
```
  40
−  3
```

14.
```
  80
−  5
```

15.
```
  20
−  1
```

16.
```
  10
−  6
```

17.
```
  30
− 15
```

18.
```
  70
−  8
```

19.
```
  60
− 39
```

20.
```
  50
−  9
```

Zeros in Subtraction

Line up the numbers. Subtract only digits with the same place value.

Use These Steps

Subtract   40 − 28

1. Line up the digits in columns.

2. Since you can't subtract 8 from 0, borrow 1 ten. Rename.

3. Subtract the ones. 10 − 8 = 2 ones. Subtract the tens. 3 − 2 = 1 ten. Check your answer.

Check:

```
                    3 10              3 10       1
    40              4 0               4 0        1 2
  − 28            − 2 8             − 2 8       +2 8
                                     1 2         4 0
```

Subtract. Use addition to check your answers.

1.
```
  5 10           1
  6 0           45
− 1 5         + 15
  4 5           60
```

2.
```
  30
−  7
```

3.
```
  40
−  9
```

4.
```
  90
− 86
```

5.   50 − 10 =

6.   70 − 11 =

7.   80 − 6 =

8.   20 − 19 =

9.   30 − 23 =

10.   60 − 50 =

11.   50 − 47 =

12.   90 − 40 =

13. Justin got a 72 on his first history test and a 90 on his second test. How much higher did he score on the second test than on the first?

14. Carmen bought a kitchen clock on sale for $19. The regular price of the clock was $30. How much money did Carmen save?

Answer _____

Answer _____

Mixed Review

Add or subtract.

1.
$$16 - 8$$

2.
$$12 - 5$$

3.
$$6 + 8$$

4.
$$15 + 23$$

5.
$$56 - 25$$

6.
$$132 - 21$$

7. $590 - 430 =$

8. $225 + 10 =$

9. $2,866 - 56 =$

10. $1,304 + 291 =$

11.
$$71 - 36$$

12.
$$55 + 49$$

13.
$$27 - 19$$

14.
$$85 + 6$$

15. $36 + 29 =$

16. $42 - 7 =$

17. $93 - 15 =$

18. $87 - 78 =$

19.
$$30 - 7$$

20.
$$90 - 65$$

21.
$$50 - 8$$

22.
$$40 + 32$$

23. $30 + 16 =$

24. $20 - 11 =$

25. $30 - 10 =$

26. $70 - 67 =$

27. The Midway bus takes 30 hours to go from Los Angeles to New Orleans. On the way, it stops in El Paso. It takes 16 hours to reach El Paso. How long does it take the bus to get from El Paso to New Orleans?

28. Another Midway bus takes 22 hours to travel from Dallas to San Diego. Then it takes 20 hours more to get to Seattle. How many hours does the whole trip take?

Answer_____

Answer_____

Brenda manages the inventory at Better Hardware. The store tries to have a set amount of stock on hand every month. Brenda has a chart of this set amount. At the end of each month, she counts the stock on hand. Then she subtracts this number from the amount on her chart so she will know how much stock to order.

The store needs to have 90 boxes of bolts in stock. Brenda counted 42 boxes. How many more boxes of bolts should Brenda order?

$$90$$
$$-42$$
$$48$$

Brenda should order 48 more boxes.

1. Look at Brenda's chart. The chart shows the number of boxes in stock at the end of the month, and the number the store needs to have in stock. How many boxes of each item should Brenda order? Complete the

	Nails	Screws	Washers	Nuts
Amount needed in stock	80	74	60	45
In stock at end of the month	78	65	52	39
Amount Brenda should order	2			

2. Last week Brenda counted the cans of paint in stock at the beginning and end of each day. She made a chart to help keep a record. On Monday morning, Brenda counted 87 cans. The store sold 7 cans of paint that day. Each day, she recorded on the chart the number of cans sold. How many cans of paint did Brenda have left at the end of each day? Complete the chart.

	Monday	Tuesday	Wednesday	Thursday	Friday	Saturday
Beginning Count	87	80				
Cans Sold	7	14	12	7	9	28
Ending Count	80					

Subtracting from Larger Numbers with Renaming

To subtract larger numbers, use the same steps you have been using for two-digit subtraction problems. You may need to rename two or more times.

Use These Steps

Subtract   367
          − 179

1. To subtract the ones, borrow 1 ten. Rename.
17 − 9 = 8 ones.

```
    5 17
  3 6̸ 7̸
 − 1 7 9
        8
```

2. To subtract the tens, borrow 1 hundred. Rename. Now there are 15 tens.
15 − 7 = 8 tens.

```
        15
  2 5̸ 17
  3 6̸ 7̸
 − 1 7 9
      8 8
```

3. Subtract the hundreds.
2 − 1 = 1 hundred.
Check your answer.

```
        15              Check:
  2 5̸ 17                 1 1
  3 6̸ 7̸                  1 8 8
 − 1 7 9               + 1 7 9
  1 8 8                  3 6 7
```

Subtract. Use addition to check your answers.

1.
```
    13
  2 3̸ 15        1 1
  3 4̸ 5̸         1 7 7
 − 1 6 8       + 1 6 8
  1 7 7         3 4 5
```

2.
```
   561
 − 373
```

3.
```
   941
 − 182
```

4.
```
   750
 − 521
```

5.
```
   981
 − 873
```

6.
```
   920
 − 892
```

7.
```
   472
 − 265
```

8.
```
   731
 − 536
```

9.
```
   483
 − 399
```

10.
```
   234
 − 159
```

11.
```
   388
 − 198
```

12.
```
   543
 − 367
```

13.
```
   635
 − 246
```

14.
```
   841
 − 632
```

15.
```
   724
 − 425
```

16.
```
   920
 − 563
```

Subtracting from Larger Numbers with Renaming

When you subtract larger numbers, be sure to line up the digits in the correct columns.

Use These Steps

Subtract 2,890 − 962

1. Line up the digits. To subtract the ones, borrow 1 ten. Rename. 10 − 2 = 8 ones.

```
    8 10
 2,8 9 0
-  9 6 2
       8
```

2. Subtract the tens. 8 − 6 = 2 tens. No borrowing is needed.

```
    8 10
 2,8 9 0
-  9 6 2
      2 8
```

3. To subtract the hundreds, borrow 1 thousand. Rename 1 thousand as 10 hundreds. 18 − 9 = 9 hundreds. Subtract the thousands. 1 − 0 = 1 thousand. Check your answer.

Check:

```
 1 18 8 10        1  1
 2,8 9 0        1,9 2 8
-  9 6 2        +  9 6 2
 1,9 2 8        2,8 9 0
```

Subtract. Use addition to check your answers.

1.
$$3,640 - 1,983 =$$

```
 15 13
 2 5 3 10      1 1 1
 3,6 4 0      1,657
- 1,983      + 1,983
 1,657       3,640
```

2.
$$4,794 - 1,887 =$$

3.
$$6,550 - 3,458 =$$

4.
$$11,832 - 4,659 =$$

5.
$$25,611 - 14,703 =$$

6.
$$37,730 - 1,982 =$$

7.
$$124,653 - 48,927 =$$

8.
$$362,141 - 183,529 =$$

9.
$$943,217 - 56,199 =$$

Mixed Review

Subtract.

1.
```
  258
- 169
```

2.
```
  961
- 382
```

3.
```
1,545
-  39
```

4.
```
  652
- 399
```

5.
$$276 - 93 =$$

6.
$$772 - 89 =$$

7.
$$2,435 - 95 =$$

8.
```
 12,490
-  1,652
```

9.
```
 53,925
- 41,799
```

10.
```
 98,324
-   990
```

11.
$$14,320 - 11,965 =$$

12.
$$25,030 - 19,019 =$$

13.
$$52,340 - 967 =$$

14.
```
 125,330
-  26,592
```

15.
```
 980,352
-  30,199
```

16.
```
 247,822
- 172,653
```

17. The diameter (distance through the center) of the smallest planet, Pluto, is 1,420 miles. The diameter of the largest planet, Jupiter, is 88,640 miles. How much larger is the diameter of Jupiter than the diameter of Pluto?

18. The diameter of Earth is 7,926 miles. The sun has a diameter of 863,027 miles. How much larger is the diameter of the sun than the diameter of Earth?

Answer _____

Answer _____

Zeros in Subtraction

Zeros may be in two places in larger numbers. When you subtract from two zeros, you will need to borrow two or more times.

Use These Steps

Subtract 700
− 375

1. There are no ones and no tens in the top number. To subtract, borrow from the hundreds column. Rename 1 hundred as 10 tens.

$$\begin{array}{r} \overset{6\ 10}{7\cancel{0}0} \\ -375 \\ \hline \end{array}$$

2. There are now 10 tens. Rename 1 ten as 10 ones.

$$\begin{array}{r} \overset{9}{\overset{6\cancel{10}10}{7\cancel{0}\cancel{0}}} \\ -375 \\ \hline \end{array}$$

3. There are now 10 ones, 9 tens, and 6 hundreds. Subtract.
   10 − 5 = 5 ones.
   9 − 7 = 2 tens.
   6 − 3 = 3 hundreds.
   Check your answer.

$$\begin{array}{r} \overset{9}{\overset{6\cancel{10}10}{7\cancel{0}\cancel{0}}} \\ -375 \\ \hline 325 \end{array}$$

Check:
$$\begin{array}{r} \overset{1\ 1}{325} \\ +375 \\ \hline 700 \end{array}$$

Subtract. Use addition to check your answers.

1.
$$\begin{array}{r} \overset{9}{\overset{4\cancel{10}10}{5\cancel{0}\cancel{0}}} \\ -263 \\ \hline 237 \end{array}$$
$$\begin{array}{r} \overset{1\ 1}{237} \\ +263 \\ \hline 500 \end{array}$$

2.
$$\begin{array}{r} 300 \\ -\ 98 \\ \hline \end{array}$$

3.
$$\begin{array}{r} 400 \\ -\ 56 \\ \hline \end{array}$$

4.
$$\begin{array}{r} 900 \\ -398 \\ \hline \end{array}$$

5.
$$\begin{array}{r} 3,100 \\ -1,642 \\ \hline \end{array}$$

6.
$$\begin{array}{r} 5,500 \\ -2,307 \\ \hline \end{array}$$

7.
$$\begin{array}{r} 2,600 \\ -\ 593 \\ \hline \end{array}$$

8.
$$\begin{array}{r} 8,700 \\ -\ 477 \\ \hline \end{array}$$

9.
$$\begin{array}{r} 18,700 \\ -17,856 \\ \hline \end{array}$$

10.
$$\begin{array}{r} 26,900 \\ -\ 9,459 \\ \hline \end{array}$$

11.
$$\begin{array}{r} 57,300 \\ -\ 3,750 \\ \hline \end{array}$$

12.
$$\begin{array}{r} 95,500 \\ -10,022 \\ \hline \end{array}$$

13.
$$\begin{array}{r} 6,200 \\ -2,341 \\ \hline \end{array}$$

14.
$$\begin{array}{r} 27,100 \\ -\ 5,248 \\ \hline \end{array}$$

15.
$$\begin{array}{r} 83,300 \\ -17,306 \\ \hline \end{array}$$

16.
$$\begin{array}{r} 41,600 \\ -\ 782 \\ \hline \end{array}$$

Zeros in Subtraction

To subtract from three or more zeros, you will need to borrow three or more times.

Use These Steps

Subtract 3,000 − 2,506

1. Line up the digits.

2. Rename three times.

3. Subtract.
10 − 6 = 4 ones.
9 − 0 = 9 tens.
9 − 5 = 4 hundreds.
2 − 2 = 0 thousands.
Check your answer.

```
                    9 9                    9 9              Check:
                  2 10 10 10             2 10 10 10          1 1 1
     3,000          3,0 0 0               3,0 0 0             4 9 4
   − 2,506        − 2,5 0 6             − 2,5 0 6           +2,5 0 6
                                           4 9 4             3,0 0 0
```

Subtract. Use addition to check your answers.

1.
```
     9 9
   4 10 10 10      1 1 1
     5,0 0 0       3,505
   − 1,4 9 5     + 1,495
     3,5 0 5       5,000
```

2.
```
     7,000
   − 6,703
```

3.
```
     1,000
   −  985
```

4.
```
     3,000
   −  770
```

5.
```
    26,000
   − 2,891
```

6.
```
    42,000
   − 15,407
```

7.
```
   196,000
   − 99,396
```

8.
```
   437,000
   − 256,930
```

9. 53,000 − 47,926 =

10. 148,000 − 129,672 =

11. 539,000 − 2,659 =

12. 290,000 − 32,619 =
```
      9 9 9
    8 10 10 10 10      1 1 1 1
    2 9 0,0 0 0        257,381
   −   3 2,6 1 9      +  32,619
    2 5 7,3 8 1        290,000
```

13. 300,000 − 56,032 =

14. 800,000 − 136,275 =

Borrowing Across Zeros

Some subtraction problems may have one or more zeros in the middle part of the top number. When this happens, you will need to borrow across the zeros.

Use These Steps

Subtract 501 − 299

1. Line up the digits. To subtract ones, borrow 1 hundred across the zero from the hundreds column. Rename 1 hundred as 10 tens.

$$\begin{array}{r} {}^{4\ 10}\\ 5\,\cancel{0}\,1 \\ -\,2\,9\,9 \\ \hline \end{array}$$

2. Borrow 1 ten and rename as 10 ones. There are now 4 hundreds, 9 tens, and 11 ones.

$$\begin{array}{r} {}^{9}\\ {}^{4\ \cancel{10}\ 11}\\ 5\,\cancel{0}\,1 \\ -\,2\,9\,9 \\ \hline \end{array}$$

3. Subtract.
   11 − 9 = 2 ones.
   9 − 9 = 0 tens.
   4 − 2 = 2 hundreds. Check your answer.

$$\begin{array}{r} {}^{9}\\ {}^{4\ \cancel{10}\ 11}\\ 5\,\cancel{0}\,1 \\ -\,2\,9\,9 \\ \hline 2\,0\,2 \end{array}\qquad \begin{array}{r} \text{Check:}\\ {}^{1\ 1}\\ 2\,0\,2 \\ +\,2\,9\,9 \\ \hline 5\,0\,1 \end{array}$$

Subtract. Use addition to check your answers.

1.
$$\begin{array}{r} {}^{9}\\ {}^{6\ \cancel{10}\ 16}\\ 7\,\cancel{0}\,6 \\ -\,2\,5\,7 \\ \hline 4\,4\,9 \end{array}\qquad \begin{array}{r} {}^{1\ 1}\\ 4\,4\,9 \\ +\,2\,5\,7 \\ \hline 7\,0\,6 \end{array}$$

2.
$$\begin{array}{r} 301 \\ -\,199 \\ \hline \end{array}$$

3.
$$\begin{array}{r} 508 \\ -\,369 \\ \hline \end{array}$$

4.
$$\begin{array}{r} 903 \\ -\,424 \\ \hline \end{array}$$

5.
$$\begin{array}{r} 207 \\ -\,78 \\ \hline \end{array}$$

6.
$$\begin{array}{r} 403 \\ -\,56 \\ \hline \end{array}$$

7.
$$\begin{array}{r} 503 \\ -\,88 \\ \hline \end{array}$$

8.
$$\begin{array}{r} 707 \\ -\,99 \\ \hline \end{array}$$

9.
$$\begin{array}{r} 1{,}603 \\ -\,348 \\ \hline \end{array}$$

10.
$$\begin{array}{r} 4{,}302 \\ -\,773 \\ \hline \end{array}$$

11.
$$\begin{array}{r} 9{,}206 \\ -\,3{,}837 \\ \hline \end{array}$$

12.
$$\begin{array}{r} 7{,}701 \\ -\,2{,}995 \\ \hline \end{array}$$

13. 406 − 249 =

14. 301 − 93 =

15. 1,702 − 635 =

Borrowing Across Zeros

To subtract across one or more zeros, you may need to borrow across the zeros from the hundreds or thousands place.

Use These Steps

Subtract 35,003 − 3,236

1. Line up the digits.

```
  35,003
−  3,236
```

2. To subtract the ones, borrow across the zeros from the thousands column. Rename. There are now 4 thousands, 9 hundreds, 9 tens, and 13 ones.

```
        9 9
     4 10 10 13
   3 5,0 0 3
 −   3,2 3 6
```

3. Subtract. Check your answer.

```
        9 9
     4 10 10 13
   3 5,0 0 3
 −   3,2 3 6
   3 1,7 6 7
```

Check:
```
     1 1 1
   3 1,7 6 7
 +   3,2 3 6
   3 5,0 0 3
```

Subtract. Use addition to check your answers.

1.
```
   9 9
 5 10 10 13
   6,0 0 3      2,516
 − 3,4 8 7    + 3,487
   2,5 1 6      6,003
```

2.
```
   4,001
 − 1,892
```

3.
```
   1,007
 −   998
```

4.
```
   2,006
 −   347
```

5. 7,002 − 685 =

6. 19,006 − 8,149 =

7. 25,009 − 7,658 =

8. 10,032 − 765 =

9. 90,026 − 1,493 =

10. 130,082 − 15,996 =

11. 3,002 − 926 =

12. 20,029 − 1,577 =

13. 100,251 − 74,844 =

Problem Solving: Using Rounding and Estimating

When you work with larger numbers, it is sometimes easier to use rounded numbers rather than exact numbers. In rounding a number to the nearest thousand, remember that if the digit in the hundreds place ends in 5, round to the next higher number in the thousands place.

Example This chart lists ten of the most famous mountains in the world and their height above sea level. About how many feet taller is Mount McKinley than Mont Blanc?

Mountain	Height in Feet	Mountain	Height in Feet
Aconcagua	22,831	Kilimanjaro	19,340
Mauna Loa	13,677	Mont Blanc	15,771
Mount Everest	29,028	Mount Fuji	12,388
Mount Kenya	17,058	Mount Logan	19,524
Mount McKinley	20,320	Mount Rainier	14,410

▶ **Step 1.** Round each number to the nearest thousand.

Mount McKinley   20,320   rounds to 20,000
Mont Blanc   15,771   rounds to 16,000

▶ **Step 2.** Compare the rounded numbers.

20,000 > 16,000, so Mount McKinley is taller than Mont Blanc.

▶ **Step 3.** Subtract the smaller number from the larger number.

$$
\begin{array}{r}
^{1\,10}\\
2\!\!\!/0,000\\
-16,000\\
\hline
4,000
\end{array}
$$

Mount McKinley is about 4,000 feet taller than Mont Blanc.

Solve.

1. Round the height of Kilimanjaro to the nearest thousand.

2. Round the height of Aconcagua to the nearest thousand.

Answer _____

Answer _____

3. Round the height of Mount Rainier to the nearest thousand.

Answer_____

4. Round the height of Mauna Loa to the nearest thousand.

Answer_____

5. Round the height of Mount Fuji to the nearest thousand.

Answer_____

6. Compare the rounded heights of Kilimanjaro and Mauna Loa. Which mountain is taller?

Answer_____

7. Compare the rounded heights of Mount Fuji and Mount McKinley. Which mountain is taller?

Answer_____

8. About how many feet taller is Mauna Loa than Mount Fuji?

Answer_____

9. About how many feet taller is Aconcagua than Mount Rainier?

Answer_____

10. About how many feet shorter than Mount Everest is Mount Logan?

Answer_____

11. About how many feet shorter than Mount McKinley is Mount Kenya?

Answer_____

12. About how many feet shorter than Kilimanjaro is Mont Blanc?

Answer_____

Unit 3 Review

Subtract. Use addition to check your answers.

1.
$$17 - 8$$

2.
$$13 - 7$$

3.
$$12 - 4$$

4.
$$27 - 13$$

5.
$$94 - 83$$

6.
$$85 - 15$$

7. $16 - 9 =$

8. $11 - 9 =$

9. $36 - 4 =$

10. $72 - 10 =$

11.
$$367 - 243$$

12.
$$935 - 25$$

13.
$$657 - 343$$

14.
$$1{,}507 - 1{,}402$$

15. $193 - 141 =$

16. $280 - 140 =$

17. $2{,}327 - 1{,}304 =$

18.
$$53 - 17$$

19.
$$26 - 9$$

20.
$$47 - 38$$

21.
$$93 - 86$$

22. $65 - 56 =$

23. $44 - 9 =$

24. $52 - 42 =$

25. $72 - 29 =$

26.
$$20 - 15$$

27.
$$30 - 29$$

28.
$$90 - 56$$

29.
$$50 - 3$$

30. $60 - 6 =$

31. $80 - 5 =$

32. $40 - 26 =$

33. $70 - 62 =$

Subtract. Use addition to check your answers.

34.
$$\begin{array}{r} 641 \\ -\ 572 \\ \hline \end{array}$$

35.
$$\begin{array}{r} 916 \\ -\ 709 \\ \hline \end{array}$$

36.
$$\begin{array}{r} 2{,}350 \\ -\ \ 291 \\ \hline \end{array}$$

37.
$$\begin{array}{r} 1{,}552 \\ -\ \ 497 \\ \hline \end{array}$$

38. $5{,}623 - 1{,}927 =$

39. $17{,}420 - 15{,}523 =$

40. $25{,}062 - 9{,}059 =$

41.
$$\begin{array}{r} 400 \\ -\ 293 \\ \hline \end{array}$$

42.
$$\begin{array}{r} 700 \\ -\ 654 \\ \hline \end{array}$$

43.
$$\begin{array}{r} 800 \\ -\ 427 \\ \hline \end{array}$$

44.
$$\begin{array}{r} 3{,}506 \\ -\ 1{,}968 \\ \hline \end{array}$$

45. $15{,}006 - 13{,}558 =$

46. $72{,}008 - 9{,}662 =$

47. $90{,}800 - 1{,}857 =$

48. $29{,}000 - 17{,}627 =$

49. $700{,}000 - 493{,}274 =$

50. $680{,}000 - 14{,}901 =$

Below is a list of the problems in this review and the pages on which the skills are taught. If you missed any problems, turn to the pages listed and practice the skills. Then correct the problems you missed in the Unit Review.

Problems	Pages	Problems	Pages
1–3, 7–8	55–58	26–33	68–69
4–6, 9–10	60–61	34–40	72–73
11–17	62	41–50	75–78
18–25	66–67		

Multiplication is the same as repeated addition. For example, if you have a case of cherry soda that has 4 six-packs in it, you could add sixes to find out how many cans of soda you have. 6 + 6 + 6 + 6 = 24 cans. It is much easier, though, to multiply 4 by 6 to find the answer.

To be successful at multiplying numbers, you must memorize the multiplication facts. These facts are the basis of every multiplication problem. In this unit, you will learn the multiplication facts and how to multiply by one-digit, two-digit, and three-digit numbers. You will also learn about multiplying with zeros.

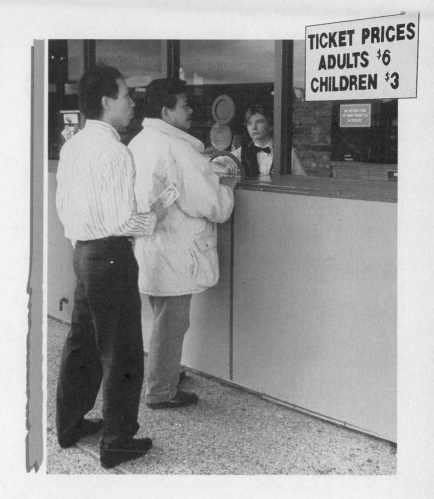

Getting Ready

You should be familiar with the skills on this page and the next before you begin this unit. To check your answers, turn to page 187.

 When you are working with whole numbers, place value is important in setting up problems, renaming, and lining up answers.

Write the value of the underlined digit in each number.

1. 3<u>2</u>   **2 ones** _____

2. <u>5</u>47 _____

3. 1,0<u>6</u>1 _____

4. <u>1</u>,495 _____

5. <u>6</u>,886 _____

6. 5,0<u>0</u>0 _____

To set up an addition problem, first line up the digits.
Then solve the problem.

Line up the digits in each problem and solve.

7.

342 + 11 =

```
  342
+  11
-----
  353
```

8.

1,365 + 32 =

9.

15,341 + 1,231 =

For review, see Unit 2, pages 36-37.

 Renaming in addition will help you learn to rename in
multiplication.

Add.

10.

297 + 48 =

```
  11
  297
+  48
-----
  345
```

11.

1,619 + 391 =

12.

83,457 + 19,543 =

For review, see Unit 2, pages 43-44.

 Zero plus any number is that number.

Add.

13.
```
  200
+ 591
-----
  791
```

14.
```
  1,709
+     3
-------
```

15.
```
  506
+  37
-----
```

16.
```
  7,006
+ 8,906
-------
```

17.
```
   20,009
+  36,304
---------
```

18.

306 + 42 =

19.

1,030 + 2,091 =

20.

600 + 5,300 =

21.

30,004 + 502 + 2,007 =

22.

10,600 + 3,400 + 20 =

For review, see Unit 2, page 45.

Multiplication Facts

To multiply larger numbers, you should first know the basic
multiplication facts. You will find it helpful to know the following facts
by heart.

**Multiply the following numbers to complete each row. Notice that the answers form a
pattern. Notice also that zero times any number equals zero.**

1.
0	1	2	3	4	5	6	7	8	9
×1	×1	×1	×1	×1	×1	×1	×1	×1	×1
0	1	2	3	4	5	6	7	8	9

2.
0	1	2	3	4	5	6	7	8	9
×2	×2	×2	×2	×2	×2	×2	×2	×2	×2

3.
0	1	2	3	4	5	6	7	8	9
×3	×3	×3	×3	×3	×3	×3	×3	×3	×3

4.
0	1	2	3	4	5	6	7	8	9
×4	×4	×4	×4	×4	×4	×4	×4	×4	×4

5.
0	1	2	3	4	5	6	7	8	9
×5	×5	×5	×5	×5	×5	×5	×5	×5	×5

6.
0	1	2	3	4	5	6	7	8	9
×6	×6	×6	×6	×6	×6	×6	×6	×6	×6

7.
0	1	2	3	4	5	6	7	8	9
×7	×7	×7	×7	×7	×7	×7	×7	×7	×7

8.
0	1	2	3	4	5	6	7	8	9
×8	×8	×8	×8	×8	×8	×8	×8	×8	×8

9.
0	1	2	3	4	5	6	7	8	9
×9	×9	×9	×9	×9	×9	×9	×9	×9	×9

Multiplication Facts Practice

Use the multiplication facts from page 85 to complete the table.

Example Find an empty box in the table. Look up the column to the top row of numbers. Then move left from the box to the number at the beginning of the row. Multiply these two numbers. Write the answer in the empty box.

×	0	1	2	3	4	5	6	7	8	9
0	0									
1		1								
2			4							
3				9						
4					16					
5						25				
6							36			
7								49		
8									64	
9										81

To use the table to complete the multiplication facts, find one number at the top of a column and the other number at the beginning of a row. The answer is the number in the box where the row and the column meet.

Complete the following multiplication facts.

1. $6 \times 7 = 42$
2. $3 \times 9 =$
3. $5 \times 7 =$
4. $2 \times 0 =$
5. $9 \times 9 =$
6. $8 \times 4 =$

7. $5 \times 6 =$
8. $3 \times 4 =$
9. $7 \times 8 =$
10. $2 \times 6 =$
11. $6 \times 6 =$
12. $3 \times 2 =$

13. $4 \times 4 =$
14. $0 \times 8 =$
15. $1 \times 5 =$
16. $9 \times 8 =$
17. $4 \times 1 =$
18. $7 \times 3 =$

19. $6 \times 9 =$
20. $8 \times 5 =$
21. $4 \times 2 =$
22. $3 \times 3 =$
23. $0 \times 1 =$
24. $2 \times 5 =$

Multiplication Facts Practice

Multiplication problems can be written vertically or horizontally.

multiplicand $\longrightarrow$ $\quad 9$
multiplier $\quad\longrightarrow$ $\underline{\times\,6}$ $\quad$ is the same as $6 \times 9 = 54$
product $\quad\longrightarrow$ $\quad 54$

Complete the following multiplication facts. You can use the table on page 86 if you need help remembering the facts.

1. $5 \times 5 = \boxed{25}$

2. $2 \times 8 = \square$

3. $7 \times 3 = \square$

4. $9 \times 4 = \square$

5. $4 \times \boxed{8} = 32$

6. $6 \times \square = 0$

7. $8 \times \square = 8$

8. $2 \times \square = 10$

9. $\square \times 6 = 24$

10. $\square \times 3 = 9$

11. $\square \times 7 = 42$

12. $\square \times 2 = 18$

13. $1 \times \square = 0$

14. $5 \times 9 = \square$

15. $\square \times 6 = 18$

16. $9 \times \square = 36$

17. $8 \times \square = 40$

18. $\square \times 4 = 12$

19. $7 \times 8 = \square$

20. $5 \times 6 = \square$

Fill in the boxes with any numbers that make the products. There may be more than one set of numbers that makes a true statement. Use the table on page 86 if you need help.

21. $\boxed{5} \times \boxed{4} = 20$

22. $\square \times \square = 16$

23. $\square \times \square = 27$

24. $\square \times \square = 32$

25. $\square \times \square = 42$

26. $\square \times \square = 12$

27. $\square \times \square = 9$

28. $\square \times \square = 0$

29. $\square \times \square = 8$

30. $\square \times \square = 54$

31. $\square \times \square = 45$

32. $\square \times \square = 64$

33. $\square \times \square = 15$

34. $\square \times \square = 21$

35. $\square \times \square = 36$

36. $\square \times \square = 72$

37. $\square \times \square = 24$

38. $\square \times \square = 25$

39. $\square \times \square = 10$

40. $\square \times \square = 18$

Real-Life Application   On the Job

You can use multiplication to help you figure out total cost when you are buying more than one of the same thing.

Barron's Office Supply   *21*

$2 per box — *Write Idea* pens
$4 each — Staplers
$2 each — Notebooks
$3 each — Calendars

For each problem below, use the prices from the catalog to work out each answer. Circle the letter next to the correct choice.

1. Carlos is ordering supplies for his company. He wants to buy 6 boxes of pens. How much will 6 boxes of pens cost?

   a. $9
   b. $12
   c. $18

2. Carlos doesn't want to spend more than $20 for pens. If he buys 6 boxes, he will spend

   a. less than $20.
   b. more than $20.
   c. exactly $20.

3. There are 9 pens in each box. If Carlos orders 6 boxes, how many pens will he get?

   a. 27
   b. 54
   c. 15

4. There are 5 offices in Carlos' company. He needs to order 1 stapler for each office. Which choice shows how to find the total cost of the staplers?

   a. 5 + $4
   b. 5 − $4
   c. 5 × $4

5. Carlos needs to order 8 notebooks. How much will 8 notebooks cost?

   a. $10
   b. $16
   c. $20

6. Carlos also wants to order 5 calendars. How much will it cost for the 5 calendars and the 8 notebooks all together?
   (Hint: Use the answer for question 5 to solve this problem.)

   a. less than $40
   b. more than $40
   c. exactly $40

Multiplying by One-Digit Numbers

When you multiply by a one-digit number, use the multiplication facts. Begin by multiplying the ones digits. Be sure to line up your answers in the correct column.

Use These Steps

Multiply
```
  12
×  4
```

1. Be sure that the digits are lined up.

```
  12
×  4
```

2. Multiply the ones. $4 \times 2 = 8$ ones. Put the 8 in the ones column.

```
  12
×  4
   8
```
← 8 ones

3. Multiply the tens. $4 \times 1 = 4$ tens. Put the 4 in the tens column.

```
  12
×  4
  48
```
← 4 tens

Multiply.

1.
```
  23
× 3
  69
```

2.
```
  11
× 2
```

3.
```
  13
× 3
```

4.
```
  24
× 2
```

5.
```
  21
× 3
```

6.
```
  11
× 3
```

7.
```
  32
× 1
```

8.
```
  41
× 2
```

9.
```
  21
× 4
```

10.
```
  42
× 2
```

11.
```
  33
× 2
```

12.
```
  22
× 4
```

13.
```
  153
×   1
```

14.
```
  243
×   2
```

15.
```
  233
×   3
```

16.
```
  432
×   2
```

17.
```
  312
×   3
```

18.
```
  211
×   4
```

19.
```
  123
×   3
```

20.
```
  231
×   2
```

21.
```
  562
×   1
```

22.
```
  403
×   2
```

23.
```
  330
×   3
```

24.
```
  120
×   4
```

25. $36 \times 1 =$
```
  36
× 1
  36
```

26. $23 \times 3 =$

27. $102 \times 4 =$

28. $344 \times 2 =$

Multiplying by One-Digit Numbers

When you multiply, write the answer from right to left, starting with the ones column.

Use These Steps

Multiply   312
         ×   4

1. Be sure that the digits are lined up. Multiply the ones. 4 × 2 = 8 ones. Put the 8 in the ones column.

2. Multiply the tens. 4 × 1 = 4 tens. Put the 4 in the tens column.

3. Multiply the hundreds. 4 × 3 = 12 hundreds. The answer is more than 10. Put the 2 in the hundreds column. Put the 1 in the thousands column.

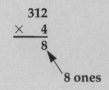
312
× 4
8

8 ones

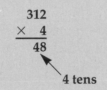
312
× 4
48

4 tens

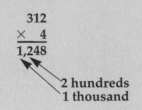
312
× 4
1,248

2 hundreds
1 thousand

Multiply.

1.
   53
   × 2
   106

2.
   41
   × 3

3.
   32
   × 4

4.
   211
   × 5

5.
   412
   × 4

6.
   422
   × 3

7.
   311
   × 7

8.
   522
   × 4

9.
   7,112
   × 3

10.
   6,113
   × 2

11. 62 × 4 =

12. 83 × 3 =

13. 71 × 7 =

14. 812 × 4 =

15. Ethel is making cookies. Her recipe will make 52 cookies. If Ethel makes 3 times the recipe, how many cookies will she bake?

16. Alejandro bought 21 sheets of plywood to build a garage. Each sheet cost $5. How much did he spend for plywood?

Answer _____

Answer _____

Multiplying with Zeros

When you multiply zeros, you will have a zero in your answer.
Remember that zero times any number is zero.

Use These Steps

Multiply   903
         × 2

1. Be sure that the digits are lined up. Multiply the ones. 2 × 3 = 6 ones. Put the 6 in the ones column.

2. Multiply the tens. 2 × 0 = 0 tens. Put the 0 in the tens column.

3. Multiply the hundreds. 2 × 9 = 18 hundreds. The answer is more than 10. Put the 8 in the hundreds column. Put the 1 in the thousands column.

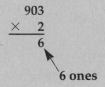

6 ones

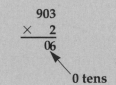

0 tens

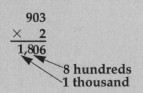

8 hundreds
1 thousand

Multiply.

1.
```
   503
×    3
 1,509
```

2.
```
    60
×    4
```

3.
```
   201
×    9
```

4.
```
    40
×    3
```

5.
```
   704
×    2
```

6.
```
 3,002
×    4
```

7.
```
 5,023
×    2
```

8.
```
 7,001
×    8
```

9.
```
 6,020
×    3
```

10.
```
 4,002
×    4
```

11. 2,001 × 7 =

12. 6,013 × 3 =

13. 5,102 × 4 =

14. 6,010 × 8 =

15. Dot's paycheck is $603 every 2 weeks. How much does she make in 4 weeks?

16. To move into her new house, Dot has to pay 3 months' rent in advance. Rent is $302. How much will she have to pay?

Answer _____

Answer _____

Mixed Review

Add, subtract, or multiply.

1.
$$\begin{array}{r} 3 \\ + 9 \\ \hline \end{array}$$

2.
$$\begin{array}{r} 7 \\ \times 6 \\ \hline \end{array}$$

3.
$$\begin{array}{r} 4 \\ \times 8 \\ \hline \end{array}$$

4.
$$\begin{array}{r} 9 \\ - 6 \\ \hline \end{array}$$

5.
$$\begin{array}{r} 18 \\ - 9 \\ \hline \end{array}$$

6.
$$\begin{array}{r} 8 \\ + 9 \\ \hline \end{array}$$

7.
$$\begin{array}{r} 5 \\ \times 6 \\ \hline \end{array}$$

8.
$$\begin{array}{r} 45 \\ - 4 \\ \hline \end{array}$$

9.
$$\begin{array}{r} 66 \\ + 3 \\ \hline \end{array}$$

10.
$$\begin{array}{r} 42 \\ \times 2 \\ \hline \end{array}$$

11.
$$\begin{array}{r} 12 \\ + 7 \\ \hline \end{array}$$

12.
$$\begin{array}{r} 31 \\ \times 3 \\ \hline \end{array}$$

13.
$$\begin{array}{r} 54 \\ - 4 \\ \hline \end{array}$$

14.
$$\begin{array}{r} 90 \\ + 9 \\ \hline \end{array}$$

15. $86 - 6 =$

16. $15 + 4 =$

17. $12 \times 4 =$

18. $63 - 2 =$

19. $20 \times 3 =$

20. $40 \times 2 =$

21. $96 + 3 =$

22. $30 \times 3 =$

23.
$$\begin{array}{r} 210 \\ \times 4 \\ \hline \end{array}$$

24.
$$\begin{array}{r} 433 \\ \times 2 \\ \hline \end{array}$$

25.
$$\begin{array}{r} 534 \\ + 3 \\ \hline \end{array}$$

26.
$$\begin{array}{r} 1,986 \\ - 6 \\ \hline \end{array}$$

27.
$$\begin{array}{r} 2,102 \\ \times 3 \\ \hline \end{array}$$

28.
$$\begin{array}{r} 6,407 \\ + 2 \\ \hline \end{array}$$

29.
$$\begin{array}{r} 30 \\ \times 5 \\ \hline \end{array}$$

30.
$$\begin{array}{r} 51 \\ \times 7 \\ \hline \end{array}$$

31.
$$\begin{array}{r} 93 \\ + 7 \\ \hline \end{array}$$

32.
$$\begin{array}{r} 3,105 \\ - 6 \\ \hline \end{array}$$

33.
$$\begin{array}{r} 2,001 \\ \times 9 \\ \hline \end{array}$$

34.
$$\begin{array}{r} 1,400 \\ + 8 \\ \hline \end{array}$$

35. $53 \times 3 =$

36. $82 + 9 =$

37. $177 - 8 =$

38. $400 \times 7 =$

39. $6,200 \times 4 =$

40. $9,860 + 50 =$

41. $8,700 - 125 =$

42. $7,000 \times 9 =$

Problem Solving: Using a Circle Graph

When you are using information from a graph, decide which facts you need to solve the problem.

Example A circle graph gives you information in pie-shaped pieces. The sections of this circle graph show the items collected by the Oak Ridge Recycling Center each week.

Use the information from the graph to find how many pounds of glass the center collects every 2 weeks.

▶ **Step 1.** Decide which facts you need from the graph.

600 pounds of glass each week

▶ **Step 2.** Decide what you need to do to solve the problem.

Since the problem asks you to find the number of pounds collected every 2 weeks, multiply 600 pounds of glass by 2.

▶ **Step 3.** Set up the problem and solve.

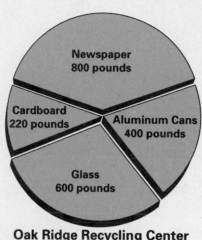

**Oak Ridge Recycling Center
Weekly Collection**

$$\begin{array}{r} 600 \\ \times\ \ 2 \\ \hline 1{,}200 \end{array}$$

The recycling center collects 1,200 pounds of glass every 2 weeks.

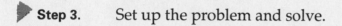

Use the information from the graph to answer the questions.

1. How many pounds of cardboard does the recycling center collect every week?

2. How many pounds of newspaper does the recycling center collect every week?

Answer_____

Answer_____

Use the information from the graph on page 93 to answer each question.

3. How many pounds of aluminum cans does the center collect every week?

Answer_____

4. How many pounds of aluminum cans does the center collect every 2 weeks?

Answer_____

5. How many pounds of aluminum cans does the center collect every month? (Hint: Use 4 weeks for 1 month.)

Answer_____

6. How many pounds of newspaper does the center collect every 3 weeks?

Answer_____

7. How many pounds of newspaper does the center collect every 2 months? (Hint: Use 8 weeks for 2 months.)

Answer_____

8. How many pounds of cardboard does the center collect every month?

Answer_____

9. How many pounds of aluminum cans and newspaper does the center collect every week?

Answer_____

10. How many pounds of aluminum cans and newspaper does the center collect every month?

Answer_____

11. How many pounds of cardboard and newspaper does the center collect every week?

Answer_____

12. How many pounds of cardboard and newspaper does the center collect every month?

Answer_____

Multiplying by Two-Digit Numbers

When you multiply by a two-digit number, multiply each digit in the top number by each digit in the bottom number. You will get two partial products. Add the partial products to get the answer.

Use These Steps

Multiply
$$\begin{array}{r} 31 \\ \times\ 42 \end{array}$$

1. Be sure that the digits are lined up. Multiply by 2 ones. $2 \times 1 = 2$ ones. Put the 2 in the ones column. $2 \times 3 = 6$ tens. Put the 6 in the tens column.

$$\begin{array}{r} 31 \\ \times\ 42 \\ \hline 62 \end{array} \leftarrow \text{partial product}$$

2. Multiply by 4 tens. $4 \times 1 = 4$ tens. Put the 4 in the tens column under the 6. $4 \times 3 = 12$ hundreds. Put the 12 to the left of the 4 tens.

$$\begin{array}{r} 31 \\ \times\ 42 \\ \hline 62 \\ 124 \end{array} \leftarrow \text{partial product}$$

3. Add the partial products to get the answer.

$$\begin{array}{r} 31 \\ \times\ 42 \\ \hline 62 \\ +124 \\ \hline 1,302 \end{array}$$

Multiply.

1.
$$\begin{array}{r} 31 \\ \times\ 22 \\ \hline 62 \\ +\ 62 \\ \hline 682 \end{array}$$

2.
$$\begin{array}{r} 12 \\ \times\ 34 \end{array}$$

3.
$$\begin{array}{r} 23 \\ \times\ 13 \end{array}$$

4.
$$\begin{array}{r} 11 \\ \times\ 56 \end{array}$$

5.
$$\begin{array}{r} 42 \\ \times\ 12 \end{array}$$

6.
$$\begin{array}{r} 23 \\ \times\ 21 \end{array}$$

7.
$$\begin{array}{r} 14 \\ \times\ 12 \end{array}$$

8.
$$\begin{array}{r} 42 \\ \times\ 22 \end{array}$$

9.
$$\begin{array}{r} 10 \\ \times\ 89 \end{array}$$

10.
$$\begin{array}{r} 12 \\ \times\ 43 \end{array}$$

11.
$$\begin{array}{r} 30 \\ \times\ 33 \end{array}$$

12.
$$\begin{array}{r} 40 \\ \times\ 21 \end{array}$$

13.
$$\begin{array}{r} 10 \\ \times\ 75 \end{array}$$

14.
$$\begin{array}{r} 20 \\ \times\ 44 \end{array}$$

15.
$$\begin{array}{r} 60 \\ \times\ 11 \end{array}$$

Multiplying by Two-Digit Numbers

When you multiply larger numbers by a two-digit number, you still get two partial products. Add the partial products to get the answer.

Use These Steps

Multiply 520 × 23

1. Line up the digits.

```
  520
×  23
```

2. Multiply by 3 ones.
Multiply by 2 tens.

```
  520
×  23
 1 560
10 40
```

3. Add the partial products.

```
   520
×   23
  1 560
+ 10 40
 11,960
```

Multiply.

1.
```
    52
×   21
    52
+ 1 04
 1,092
```

2.
```
   73
×  13
```

3.
```
   40
×  35
```

4.
```
  211
×  15
```

5.
```
  800
×  27
```

6. 602 × 43 =

7. 320 × 34 =

8. 900 × 56 =

9. 602 × 33 =

10.
```
 5,021
×   34
```

11.
```
 6,234
×   22
```

12.
```
 43,001
×    31
```

13.
```
 80,421
×    12
```

14. 3,012 × 44 =

15. 70,022 × 32 =

16. 51,000 × 97 =

17. 90,301 × 22 =

Multiplying by Three-Digit Numbers

When you multiply by a three-digit number, multiply each digit in the top number by each digit in the bottom number. You will get three partial products. Add all three partial products to get the answer.

Use These Steps

Multiply
$$\begin{array}{r} 120 \\ \times\,321 \end{array}$$

1. Be sure that the digits are lined up.

$$\begin{array}{r} 120 \\ \times\,321 \end{array}$$

2. Multiply by 1 one.
$1 \times 120 = 120$.
Multiply by 2 tens.
$2 \times 120 = 240$.
Multiply by 3 hundreds.
$3 \times 120 = 360$.

$$\begin{array}{r} 120 \\ \times\,321 \\ \hline 120 \\ 2\,40 \\ 36\,0 \end{array}$$

3. Add the partial products to get the answer.

$$\begin{array}{r} 120 \\ \times\,321 \\ \hline 120 \\ 2\,40 \\ +\,36\,0 \\ \hline 38{,}520 \end{array}$$

Multiply.

1.
$$\begin{array}{r} 312 \\ \times\,312 \\ \hline 624 \\ 3\,12 \\ +\,93\,6 \\ \hline 97{,}344 \end{array}$$

2.
$$\begin{array}{r} 430 \\ \times\,112 \end{array}$$

3.
$$\begin{array}{r} 202 \\ \times\,133 \end{array}$$

4.
$$\begin{array}{r} 301 \\ \times\,221 \end{array}$$

5.
$$\begin{array}{r} 401 \\ \times\,212 \end{array}$$

6.
$$\begin{array}{r} 240 \\ \times\,121 \end{array}$$

7.
$$\begin{array}{r} 331 \\ \times\,213 \end{array}$$

8.
$$\begin{array}{r} 100 \\ \times\,434 \end{array}$$

9.
$$\begin{array}{r} 602 \\ \times\,111 \end{array}$$

10.
$$\begin{array}{r} 400 \\ \times\,211 \end{array}$$

11.
$$\begin{array}{r} 302 \\ \times\,323 \end{array}$$

12.
$$\begin{array}{r} 200 \\ \times\,413 \end{array}$$

Multiplying by Three-Digit Numbers

When you multiply each digit in the top number by each digit in the bottom number, be sure to write the partial products in the correct columns.

Use These Steps

Multiply 321 × 124

1. Line up the digits.

$$\begin{array}{r} 321 \\ \times\ 124 \\ \hline \end{array}$$

2. Multiply by 4 ones.
   Multiply by 2 tens.
   Multiply by 1 hundred.

$$\begin{array}{r} 321 \\ \times\ 124 \\ \hline 1\ 284 \\ 6\ 42 \\ 32\ 1 \end{array}$$

3. Add the partial products.

$$\begin{array}{r} 321 \\ \times\ 124 \\ \hline 1\ 284 \\ 6\ 42 \\ +\ 32\ 1 \\ \hline 39,804 \end{array}$$

Multiply.

1.
$$\begin{array}{r} 530 \\ \times\ 231 \\ \hline 530 \\ 15\ 90 \\ +\ 106\ 0 \\ \hline 122,430 \end{array}$$

2.
$$\begin{array}{r} 600 \\ \times\ 325 \\ \hline \end{array}$$

3.
$$\begin{array}{r} 4,122 \\ \times\ \ \ 124 \\ \hline \end{array}$$

4.
$$\begin{array}{r} 91,000 \\ \times\ \ \ \ \ 753 \\ \hline \end{array}$$

5. $201 \times 564 =$

6. $4,100 \times 628 =$

7. $31,011 \times 739 =$

8. $80,110 \times 692 =$

9. Alicia owns a typing service. Alicia's employees typed 112 pages each day for 111 days. How many total pages did they type?

10. If Alicia charges $2 per page, how much did her company earn for typing the 112 pages?

Answer_____

Answer_____

Real-Life Application

When you are taking inventory, using multiplication can make your job easier. For example, most grocery store items come to the warehouse in cases. When you take inventory, you count the number of cases. Then you multiply by the number of items in the case to get a total.

Example Albert is the Freshway Grocery warehouse supervisor. He is taking inventory of paper goods. He counted 2,031 cases of paper towels. There are 12 rolls of towels to a case. How many rolls of paper towels are there in the warehouse?

Set up the problem. Then multiply.

$$
\begin{array}{r}
2{,}031 \\
\times\ \ \ \ 12 \\
\hline
4\ 062 \\
+\ 20\ 31\ \ \\
\hline
24{,}372
\end{array}
$$

There are 24,372 rolls of paper towels in the Freshway warehouse.

Solve.

1. Albert counted cases of tissues. There are 25 boxes of tissues in a case. He counted 3,001 cases. How many boxes of tissues are there in the warehouse?

   Answer_____

2. Albert counted 4,010 cases of paper plates. There are 45 packages of paper plates to the case. How many packages of paper plates are there in the warehouse?

   Answer_____

3. In the store, Pete counted rolls of paper towels. He counted 10 shelves of paper towels. There were 36 rolls on each shelf. How many rolls of paper towels are there?

   Answer_____

4. Pete counted 402 boxes of tissues on each of 3 shelves in the store. How many boxes of tissues are there in the store?

   Answer_____

Mixed Review

Add, subtract, or multiply.

1. $6 \times 70 =$

2. $5 + 8 =$

3. $12 \times 4 =$

4. $10 - 9 =$

5. $20 \times 33 =$

6. $56 - 15 =$

7. $382 + 7 =$

8. $122 \times 4 =$

9. $321 \times 4 =$

10. $923 \times 23 =$

11. $457 + 9 =$

12. $660 - 50 =$

13. $509 + 625 =$

14. $3,120 \times 41 =$

15. $9,002 + 998 =$

16. $7,051 - 151 =$

17. $801 \times 49 =$

18. $900 \times 32 =$

19. $6,000 + 7,000 =$

20. $1,000 - 500 =$

21. $5,102 \times 4 =$

22. $297 + 3 =$

23. $7,802 - 802 =$

24. $4,000 \times 17 =$

25. $1,022 \times 301 =$

26. $603 + 977 =$

27. $490 - 376 =$

28. $3,011 \times 509 =$

Multiplying by 10, 100, and 1,000

When you multiply by 10, 100, or 1,000, you don't have to write a row of partial products with zeros. There is an easier way to work these problems.

Use These Steps

Multiply 3,471
$\times$ 10

1. Multiply by 0 ones. Write the zero in the ones column.

2. Multiply by 1. Write the answer to the left of the zero. The answer is the same as the top number plus one zero.

$$\begin{array}{r} 3,471 \\ \times \quad 10 \\ \hline 0 \end{array}$$

$$\begin{array}{r} 3,471 \\ \times \quad 10 \\ \hline 34,710 \end{array}$$

Multiply.

1.
$$\begin{array}{r} 96 \\ \times 10 \\ \hline 960 \end{array}$$

2.
$$\begin{array}{r} 520 \\ \times 10 \\ \hline \end{array}$$

3.
$$\begin{array}{r} 1,085 \\ \times \quad 10 \\ \hline \end{array}$$

4.
$$\begin{array}{r} 7,831 \\ \times \quad 10 \\ \hline \end{array}$$

5.
$$\begin{array}{r} 76,300 \\ \times \quad 10 \\ \hline \end{array}$$

6.
$$\begin{array}{r} 461 \\ \times 100 \\ \hline 46,100 \end{array}$$

7.
$$\begin{array}{r} 982 \\ \times 100 \\ \hline \end{array}$$

8.
$$\begin{array}{r} 3,305 \\ \times \quad 100 \\ \hline \end{array}$$

9.
$$\begin{array}{r} 46,720 \\ \times \quad 100 \\ \hline \end{array}$$

10.
$$\begin{array}{r} 39,900 \\ \times \quad 100 \\ \hline \end{array}$$

11.
$$\begin{array}{r} 3,382 \\ \times 1,000 \\ \hline 3,382,000 \end{array}$$

12.
$$\begin{array}{r} 1,590 \\ \times 1,000 \\ \hline \end{array}$$

13.
$$\begin{array}{r} 2,706 \\ \times 1,000 \\ \hline \end{array}$$

14.
$$\begin{array}{r} 25,400 \\ \times \quad 1,000 \\ \hline \end{array}$$

15.
$$\begin{array}{r} 17,000 \\ \times \quad 1,000 \\ \hline \end{array}$$

16. $546 \times 10 =$

17. $2,389 \times 10 =$

18. $477 \times 100 =$

19. $12,903 \times 100 =$

20. $29 \times 100 =$

21. $36 \times 1,000$

22. $1,109 \times 1,000 =$

23. $20,000 \times 1,000 =$

Multiplying by 10, 100, and 1,000

When you multiply by 10, 100, or 1,000, you can figure out the answers without working them out on paper. Remember to use a comma.

Use These Steps

Multiply   297 × 100

1. Write the number you started with, 297.

2. Count the number of zeros in 100. There are 2. Put 2 zeros after 297.

297

$297 \times 100 = 29,700$

Multiply.

1. $27 \times 10 = \mathbf{270}$

2. $195 \times 10 =$

3. $3,402 \times 10 =$

4. $31 \times 100 = \mathbf{3,100}$

5. $286 \times 100 =$

6. $15,029 \times 100 =$

7. $490 \times 1,000 = \mathbf{490,000}$

8. $1,830 \times 1,000 =$

9. $27,600 \times 1,000 =$

10. $2,101 \times 1,000 =$

11. $133 \times 100 =$

12. $6,997 \times 10 =$

13. $41 \times 10 =$

14. $201 \times 100 =$

15. $33,032 \times 1,000 =$

16. $20 \times 100 =$

17. $3,100 \times 100 =$

18. $21,000 \times 1,000 =$

19. $89 \times 10 =$

20. $420 \times 10 =$

21. $5,100 \times 10 =$

22. $36 \times 1,000 =$

23. $500 \times 1,000 =$

24. $99 \times 100 =$

Problem Solving: Using Pictographs

A *pictograph* shows information by using symbols or pictures. Each picture stands for a certain number. That number is found in the *key*.

This pictograph shows some of the leading potato–growing countries in the world. The pictures stand for the number of bags of potatoes produced in one year.

Example   How many million bags of potatoes does the United States produce in one year?

▶ **Step 1.**   Count each ![potato] . There are two.

▶ **Step 2.**   One ![potato] equals 200 million bags of potatoes. To find the number of bags for two ![potato] , multiply 200 million by 2.

200 million × 2 = 400 million

Leading Potato–Growing Countries

Poland	🥔 🥔 🥔 🥔
Germany	🥔 🥔 🥔
United States	🥔 🥔
France	🥔

Key: One ![potato] equals 200 million bags of potatoes.

The United States produces 400 million bags of potatoes in one year.

Solve.

1. How many million bags of potatoes does Germany produce in one year?

2. How many million bags of potatoes does Poland produce in one year?

Answer_____

Answer_____

3. How many million bags of potatoes does France produce in one year?

4. How many million bags of potatoes does the United States produce in two years?
(Hint: Multiply the number of bags by 2.)

Answer_____

Answer_____

This pictograph shows the leading rice–growing countries in the world. The pictures stand for the number of tons of rice grown in one year.

Leading Rice–Growing Countries

China	🌾🌾🌾🌾🌾🌾🌾🌾🌾🌾
India	🌾🌾🌾🌾🌾
Indonesia	🌾🌾🌾
Bangladesh	🌾🌾
Thailand	🌾

Key: One 🌾 equals 20 million tons of rice.

Solve.

1. How many million tons of rice does India grow in one year?

   Answer_____

2. How many million tons of rice does Bangladesh grow in one year?

   Answer_____

3. How many million tons of rice does Indonesia grow in one year?

   Answer_____

4. There are 2,000 pounds in one ton. How many million pounds of rice does Thailand grow in one year? (Hint: Multiply the number of tons by 2,000.)

   Answer_____

5. How many million tons of rice does Indonesia grow in two years?

   Answer_____

6. How many million tons of rice does India grow in five years?

   Answer_____

Multiplying by One-Digit Numbers with Renaming

When you multiply two digits, the product is sometimes 10 or more. As in addition, when an answer is 10 or more, rename by carrying to the next column.

Use These Steps

Multiply
$$\begin{array}{r} 35 \\ \times\ 5 \\ \hline \end{array}$$

1. Be sure the digits are lined up.

$$\begin{array}{r} 35 \\ \times\ 5 \\ \hline \end{array}$$

2. Multiply the 5 by 5 ones. $5 \times 5 = 25$ ones. Rename 25 as 2 tens and 5 ones. Put the 5 in the ones column. Carry 2 tens to the top of the next column.

$$\begin{array}{r} {}^{2} \\ 35 \\ \times\ 5 \\ \hline 5 \end{array}$$

3. Multiply the 3 by 5 ones. $5 \times 3 = 15$ tens. Then add the carried 2 tens. $15 + 2 = 17$ tens. Write 17.

$$\begin{array}{r} {}^{2} \\ 35 \\ \times\ 5 \\ \hline 175 \end{array}$$

Multiply.

1.
$$\begin{array}{r} {}^{2} \\ 47 \\ \times\ 3 \\ \hline 141 \end{array}$$

2.
$$\begin{array}{r} 32 \\ \times\ 8 \\ \hline \end{array}$$

3.
$$\begin{array}{r} 93 \\ \times\ 7 \\ \hline \end{array}$$

4.
$$\begin{array}{r} 45 \\ \times\ 4 \\ \hline \end{array}$$

5.
$$\begin{array}{r} 78 \\ \times\ 9 \\ \hline \end{array}$$

6.
$$\begin{array}{r} 54 \\ \times\ 6 \\ \hline \end{array}$$

7.
$$\begin{array}{r} 62 \\ \times\ 5 \\ \hline \end{array}$$

8.
$$\begin{array}{r} 94 \\ \times\ 3 \\ \hline \end{array}$$

9.
$$\begin{array}{r} 57 \\ \times\ 7 \\ \hline \end{array}$$

10.
$$\begin{array}{r} 82 \\ \times\ 9 \\ \hline \end{array}$$

11.
$$\begin{array}{r} 250 \\ \times\ 4 \\ \hline \end{array}$$

12.
$$\begin{array}{r} 175 \\ \times\ 6 \\ \hline \end{array}$$

13.
$$\begin{array}{r} 340 \\ \times\ 5 \\ \hline \end{array}$$

14.
$$\begin{array}{r} 625 \\ \times\ 4 \\ \hline \end{array}$$

15.
$$\begin{array}{r} 875 \\ \times\ 9 \\ \hline \end{array}$$

16.
$$\begin{array}{r} 916 \\ \times\ 7 \\ \hline \end{array}$$

17.
$$\begin{array}{r} 432 \\ \times\ 8 \\ \hline \end{array}$$

18.
$$\begin{array}{r} 520 \\ \times\ 5 \\ \hline \end{array}$$

19.
$$\begin{array}{r} 899 \\ \times\ 3 \\ \hline \end{array}$$

20.
$$\begin{array}{r} 770 \\ \times\ 2 \\ \hline \end{array}$$

Multiplying by One-Digit Numbers with Renaming

When the product of two digits is 10 or more, rename by carrying to the next column. Multiply first, and then add carried numbers. You may need to rename several times.

Use These Steps

Multiply 4,769 × 3

1. Line up the digits.

```
  4,769
×     3
```

2. Multiply by 3 ones.

```
    2          22         222        222
  4,769      4,769      4,769      4,769
×     3    ×     3    ×     3    ×     3
      7         07        307     14,307
```

Multiply.

1.
```
 3 15
 5,416
×    9
48,744
```

2.
```
 7,328
×    6
```

3.
```
 4,465
×    4
```

4.
```
 9,270
×    3
```

5.
```
26,347
×    2
```

6.
```
52,180
×    5
```

7.
```
39,400
×    7
```

8.
```
92,000
×    9
```

9. $3,362 \times 4 =$

10. $11,900 \times 8 =$

11. $76,543 \times 2 =$

12. $6,989 \times 6 =$

13. Kay makes $24,500 each year as an office manager. She has worked for 3 years at this salary. How much money has she earned all together in the last 3 years?

14. Kay got a raise of $1,200 this year. How much will she earn after 2 years at her new salary?
(Hint: You need to add before you multiply.)

Answer_____

Answer_____

Multiplying with Zeros

When you multiply, there will sometimes be a zero in the top number. Remember that zero times any number is always zero. Don't forget to add carried numbers.

Use These Steps

Multiply   403
         × 7

1. Multiply the 3 by 7 ones. $3 \times 7 = 21$. Carry 2 tens.

$$\begin{array}{r} 2 \\ 403 \\ \times\ 7 \\ \hline 1 \end{array}$$

2. Multiply the 0 by 7 ones. $0 \times 7 = 0$. Add the carried 2.

$$\begin{array}{r} 2 \\ 403 \\ \times\ 7 \\ \hline 21 \end{array}$$

3. Multiply the 4 by 7 ones. $4 \times 7 = 28$. Write 28.

$$\begin{array}{r} 2 \\ 403 \\ \times\ 7 \\ \hline 2,821 \end{array}$$

Multiply.

1.
$$\begin{array}{r} 5 \\ 509 \\ \times\ 6 \\ \hline 3,054 \end{array}$$

2.
$$\begin{array}{r} 307 \\ \times\ 5 \\ \hline \end{array}$$

3.
$$\begin{array}{r} 906 \\ \times\ 8 \\ \hline \end{array}$$

4.
$$\begin{array}{r} 205 \\ \times\ 7 \\ \hline \end{array}$$

5.
$$\begin{array}{r} 404 \\ \times\ 4 \\ \hline \end{array}$$

6.
$$\begin{array}{r} 702 \\ \times\ 9 \\ \hline \end{array}$$

7.
$$\begin{array}{r} 907 \\ \times\ 3 \\ \hline \end{array}$$

8.
$$\begin{array}{r} 609 \\ \times\ 5 \\ \hline \end{array}$$

9.
$$\begin{array}{r} 506 \\ \times\ 6 \\ \hline \end{array}$$

10.
$$\begin{array}{r} 302 \\ \times\ 8 \\ \hline \end{array}$$

11. $703 \times 4 =$

12. $208 \times 9 =$

13. $805 \times 8 =$

14. $605 \times 5 =$

15. $808 \times 8 =$

16. $903 \times 6 =$

17. $707 \times 7 =$

18. $309 \times 9 =$

19. $504 \times 2 =$

20. $601 \times 3 =$

21. $809 \times 4 =$

22. $109 \times 8 =$

Multiplying with Zeros

Multiply first, and then add carried numbers.

Use These Steps

Multiply 7,094 × 8

1. Line up the digits.

2. Multiply by 8 ones.
   Add the carried numbers.

$$
\begin{array}{r}
7,094 \\
\times\quad 8 \\
\hline
\end{array}
\qquad
\begin{array}{r}
{}^{3} \\
7,094 \\
\times\quad 8 \\
\hline
2
\end{array}
\qquad
\begin{array}{r}
{}^{7\,3} \\
7,094 \\
\times\quad 8 \\
\hline
52
\end{array}
\qquad
\begin{array}{r}
{}^{7\,3} \\
7,094 \\
\times\quad 8 \\
\hline
752
\end{array}
\qquad
\begin{array}{r}
{}^{7\,3} \\
7,094 \\
\times\quad 8 \\
\hline
56,752
\end{array}
$$

Multiply.

1.
$$
\begin{array}{r}
{}^{4\ \ 4} \\
4,707 \\
\times\quad 6 \\
\hline
28,242
\end{array}
$$

2.
$$
\begin{array}{r}
9,073 \\
\times\quad 3 \\
\hline
\end{array}
$$

3.
$$
\begin{array}{r}
8,089 \\
\times\quad 5 \\
\hline
\end{array}
$$

4.
$$
\begin{array}{r}
3,605 \\
\times\quad 7 \\
\hline
\end{array}
$$

5.
$$
\begin{array}{r}
10,702 \\
\times\quad 8 \\
\hline
\end{array}
$$

6.
$$
\begin{array}{r}
66,095 \\
\times\quad 4 \\
\hline
\end{array}
$$

7.
$$
\begin{array}{r}
80,096 \\
\times\quad 2 \\
\hline
\end{array}
$$

8.
$$
\begin{array}{r}
90,909 \\
\times\quad 9 \\
\hline
\end{array}
$$

9. $42,008 \times 5 =$

10. $30,506 \times 7 =$

11. $20,005 \times 9 =$

12. Juanita collects the rent for the Sunnyvale Apartments. There are 6 apartments on each floor. She collects $405 per month for each apartment. How much rent does she collect from each floor every month?

13. If there are 10 floors in the building, how much rent in all does Juanita collect each month?

Answer _____

Answer _____

Mixed Review

Add, subtract, or multiply.

1.
```
    7
  × 6
```

2.
```
   17
  + 3
```

3.
```
   85
 − 13
```

4.
```
  131
 ×  3
```

5.
```
  260
 −  1
```

6.
```
  582
 +  8
```

7.
```
  603
 ×  2
```

8.
```
  478
 −  6
```

9.
```
  930
 +  6
```

10.
```
 9,016
 ×   4
```

11.
```
 70,005
 ×    9
```

12.
```
   25
 ×  4
```

13.
```
   62
 ×  8
```

14.
```
   59
 +  6
```

15.
```
   88
 +  2
```

16.
```
   93
 −  7
```

17.
```
   46
 −  9
```

18.
```
   31
 × 24
```

19.
```
   53
 + 87
```

20.
```
   63
 − 19
```

21.
```
   56
 × 11
```

22.
```
   21
 × 59
```

23.
```
   91
 − 89
```

24.
```
  412
 + 98
```

25.
```
  521
 × 16
```

26.
```
  752
 − 59
```

27.
```
 3,220
 ×  27
```

28.
```
 5,867
 −  77
```

29.
```
  782
 − 394
```

30.
```
  812
 × 422
```

31.
```
  660
 − 570
```

32.
```
 4,123
 × 313
```

33.
```
 1,575
 + 225
```

34.
```
  220
 × 10
```

35.
```
 1,012
 × 302
```

36.
```
 1,400
 − 296
```

37.
```
 5,032
 × 100
```

38.
```
 7,301
 × 203
```

Multiplying by Two-Digit Numbers with Renaming

When you multiply by a two-digit number, first multiply by the ones digit. Then multiply by the tens digit. When a product is 10 or more, rename by carrying to the next column.

Use These Steps

Multiply
```
  35
× 29
```

1. Be sure that the digits are lined up. Multiply by 9 ones. $9 \times 5 = 45$ ones. Carry the 4. $9 \times 3 = 27$ tens. Add the carried 4.

```
   4
  35
× 29
 315
```

2. Multiply by 2 tens. $2 \times 5 = 10$ tens. Carry the 1. $2 \times 3 = 6$ hundreds. Add the carried 1.

```
   1
  35
× 29
 315
  70
```

3. Add the partial products.

```
  35
× 29
 315
+ 70
1,015
```

Multiply.

1.
```
    46
  × 35
   230
+ 1 38
 1,610
```

2.
```
   59
 × 24
```

3.
```
   75
 × 53
```

4.
```
   62
 × 49
```

5.
```
   86
 × 75
```

6.
```
  234
×  57
```

7.
```
  603
×  78
```

8.
```
  574
×  39
```

9.
```
  409
×  68
```

10.
```
  775
×  23
```

11.
```
  802
×  99
```

12.
```
  625
×  44
```

13.
```
  306
×  67
```

14.
```
  504
×  29
```

15.
```
  207
×  33
```

Multiplying by Two-Digit Numbers with Renaming

When the product of two digits is 10 or more, rename by carrying to the next column. You may need to carry several times.

Use These Steps

Multiply   2,030 × 45

1. Line up the digits.

```
  2,030
×    45
```

**2. Multiply by 5 ones.
Multiply by 4 tens.**

```
  2,030
×    45
 10 150
 81 20
```

3. Add the partial products.

```
  2,030
×    45
 10 150
+ 81 20
 91,350
```

Multiply.

1.

5,316 × 25 =

```
  5,316
×    25
 26 580
+ 106 32
 132,900
```

2.

4,069 × 18 =

3.

9,208 × 45 =

4.

15,037 × 72 =

5.

24,005 × 37 =

6.

80,067 × 93 =

7. At the equator, the distance around Earth is about 24,902 miles. If a train could travel this distance 45 times, how many miles would it cover?

8. The distance from Earth to the moon and back is 477,720 miles. If the space shuttle goes to the moon and back 5 times, how many miles does it travel?

Answer _____

Answer _____

Multiplying by Three-Digit Numbers with Renaming

When you multiply by a three-digit number, you get three partial products. You may need to rename by carrying several times. Be sure to add the carried numbers.

Use These Steps

Multiply   516
           × 312

1. Be sure that the digits are lined up.

```
  516
× 312
```

2. Multiply by 2 ones.
   Multiply by 1 ten.
   Multiply by 3 hundreds.

```
   516
 × 312
 1 032
 5 16
154 8
```

3. Add the partial products.

```
    516
  × 312
  1 032
  5 16
+ 154 8
160,992
```

Multiply.

1.
```
    270
  × 483
    810
  21 60
+ 108 0
130,410
```

2.
```
  486
× 132
```

3.
```
  895
× 546
```

4.
```
  922
× 317
```

5.
```
  771
× 439
```

6.
```
  1,397
×   262
```

7.
```
  5,926
×   453
```

8.
```
 13,311
×    899
```

9.
```
 25,658
×    581
```

10.
```
 93,427
×    316
```

11. $450 \times 185 =$

12. $918 \times 576 =$

13. $2,820 \times 193 =$

14. $17,462 \times 761 =$

Multiplying by Three-Digit Numbers with Renaming

When the product of two digits is 10 or more, rename by carrying to the next column. You may need to rename several times. Remember to add the carried numbers when multiplying zeros.

Use These Steps

Multiply 409
        × 365

1. Be sure that the digits are lined up.

```
  409
× 365
```

2. Multiply by 5 ones.
Multiply by 6 tens.
Multiply by 3 hundreds.

```
   409
 × 365
 2 045
 24 54
122 7
```

3. Add the partial products.

```
     409
   × 365
   2 045
   24 54
 + 122 7
 149,285
```

Multiply.

1.
```
    506
  × 432
  1 012
  15 18
+ 202 4
218,592
```

2.
```
  208
× 593
```

3.
```
  709
× 946
```

4.
```
  405
× 127
```

5.
```
  302
× 279
```

6.
```
1,082
× 743
```

7.
```
5,070
× 629
```

8.
```
8,603
× 558
```

9.
```
2,009
× 417
```

10.
```
7,730
× 256
```

11.
```
10,340
×  145
```

12.
```
25,007
×  622
```

13.
```
50,603
×  254
```

14.
```
70,005
×  437
```

15.
```
62,400
×  618
```

Multiplying with Zeros

When there are one or more zeros in the bottom number, you can use this shortcut. Put a zero in the partial product. Put the next partial product to the left of the zero.

Use These Steps

Multiply 423 × 207

1. Line up the digits.

2. Multiply by 7 ones. Multiply by 0 tens, and put a zero in the tens column. Multiply by 2 hundreds, and put this partial product to the left of the zero.

3. Add the partial products.

$$
\begin{array}{r}
423 \\
\times\ 207 \\
\end{array}
$$

$$
\begin{array}{r}
423 \\
\times\ 207 \\
\hline
2\,961 \\
\end{array}
$$
partial product → 84 60 ← partial product

$$
\begin{array}{r}
423 \\
\times\ 207 \\
\hline
2\,961 \\
+\ 84\,60 \\
\hline
87{,}561 \\
\end{array}
$$

Multiply.

1.
$$
\begin{array}{r}
569 \\
\times\ 309 \\
\hline
5\,121 \\
+\ 170\,70 \\
\hline
175{,}821 \\
\end{array}
$$

2.
$$
\begin{array}{r}
640 \\
\times\ 605 \\
\end{array}
$$

3.
$$
\begin{array}{r}
872 \\
\times\ 504 \\
\end{array}
$$

4.
$$
\begin{array}{r}
330 \\
\times\ 708 \\
\end{array}
$$

5.
$$
\begin{array}{r}
400 \\
\times\ 307 \\
\end{array}
$$

6.
$$
\begin{array}{r}
3{,}040 \\
\times\ \ \ 900 \\
\hline
2{,}736{,}000 \\
\end{array}
$$

7.
$$
\begin{array}{r}
7{,}300 \\
\times\ \ \ 400 \\
\end{array}
$$

8.
$$
\begin{array}{r}
5{,}688 \\
\times\ \ \ 300 \\
\end{array}
$$

9.
$$
\begin{array}{r}
1{,}402 \\
\times\ \ \ 700 \\
\end{array}
$$

10.
$$
\begin{array}{r}
9{,}000 \\
\times\ \ \ 600 \\
\end{array}
$$

11. $13{,}025 \times 500 =$

12. $50{,}236 \times 906 =$

13. $72{,}000 \times 403 =$

14. $10{,}006 \times 502 =$

Multiplying with Zeros

When the number you are multiplying by has a zero, be sure to include zero in the partial product.

Multiply.

1.
```
      629
  ×   302
    1 258
 + 188 70
  189,958
```

2.
```
      583
  ×   500
```

3.
```
      641
  ×   106
```

4.
```
      870
  ×   401
```

5.
```
    1,409
  ×   200
```

6.
```
    5,664
  ×   704
```

7.
```
    8,306
  ×   800
```

8.
```
    9,200
  ×   907
```

9.
```
   43,099
  ×   102
```

10.
```
   54,320
  ×   300
```

11.
```
   89,003
  ×   705
```

12.
```
   17,000
  ×   400
```

Problem Solving: Using Rounding and Estimating

Multiplying rounded numbers is one way to estimate monthly and yearly income and expenses.

Example Mike makes $291 each week. About how much does he make in 1 year?

365 days	=	1 year
52 weeks	=	1 year
12 months	=	1 year
30 days	=	1 month
4 weeks	=	1 month

▶ **Step 1.** In the table, find how many weeks are in 1 year.

52 weeks = 1 year

▶ **Step 2.** Round Mike's weekly salary and the number of weeks in a year to the lead digit.

$291 rounds up to $300
52 rounds down to 50

▶ **Step 3.** Multiply the rounded numbers.

$$\begin{array}{r} \$300 \\ \times\ \ \ 50 \\ \hline \$15,000 \end{array}$$

Mike makes about $15,000 in 1 year.

Solve by rounding to the lead digit and then multiplying.

1. Estella makes $385 each week. About how much does she make in 1 year?

2. Lisa spends $415 per month on rent. About how much does she spend on rent in 1 year?

Answer_____

Answer_____

3. Eric's car payment is $107 each month. About how much does he spend on car payments in 1 year?

4. Harris spends $68 per month for bus fare. About how much does he spend in 1 year?

Answer_____

Answer_____

Solve.

5. Irving's car insurance costs $47 per month. About how much does he spend on car insurance in 1 year?

6. Mickey pays $12 each week for cooking lessons. About how much does he pay for the lessons in 1 year?

Answer_____

Answer_____

7. Ed spends about $11 for gas each week. About how much does he spend on gas in 1 month?

8. Loni spends about $50 each week on food. About how much does she spend on food in 1 year?

Answer_____

Answer_____

9. Art goes to a community college. He pays $110 each month for tuition and books. About how much does he pay for tuition and books in 1 year?

10. Sonia is saving $20 a week to buy a couch that costs $600. If she saves for 6 months, will she have enough money to buy the couch?
    (Hint: Use 24 weeks for 6 months.)

Answer_____

Answer_____

Unit 4 Review

Multiply.

1.
$$9 \times 9$$

2.
$$5 \times 7$$

3.
$$9 \times 8$$

4.
$$6 \times 4$$

5.
$$8 \times 3$$

6.
$$3 \times 7$$

7.
$$5 \times 8$$

8.
$$9 \times 0$$

9.
$$23 \times 2$$

10.
$$40 \times 2$$

11.
$$82 \times 4$$

12.
$$12 \times 4$$

13. $63 \times 3 =$

14. $221 \times 3 =$

15. $411 \times 7 =$

16. $71 \times 6 =$

17.
$$30 \times 6$$

18.
$$703 \times 3$$

19.
$$402 \times 5$$

20.
$$7,001 \times 9$$

21.
$$6,200 \times 4$$

22.
$$3,014 \times 2$$

23.
$$42 \times 12$$

24.
$$20 \times 34$$

25.
$$241 \times 21$$

26.
$$302 \times 14$$

27.
$$500 \times 39$$

28.
$$9,021 \times 43$$

29. $8,312 \times 33 =$

30. $1,010 \times 96 =$

31. $23,312 \times 32 =$

32. $80,001 \times 56 =$

Multiply.

33.
$$340 \times 122$$

34.
$$100 \times 567$$

35.
$$612 \times 343$$

36.
$$7{,}021 \times 422$$

37.
$$61{,}000 \times 759$$

38.
$$47 \times 10$$

39.
$$189 \times 10$$

40.
$$157 \times 100$$

41.
$$2{,}890 \times 100$$

42.
$$2{,}586 \times 1{,}000$$

43.
$$10{,}000 \times 1{,}000$$

44. $36 \times 100 =$

45. $295 \times 1{,}000 =$

46. $25 \times 10 =$

47. $3{,}406 \times 100 =$

48.
$$89 \times 3$$

49.
$$227 \times 9$$

50.
$$6{,}420 \times 7$$

51.
$$11{,}650 \times 4$$

52.
$$3{,}407 \times 2$$

53.
$$15{,}009 \times 6$$

54. $40{,}077 \times 8 =$

55. $6{,}020 \times 7 =$

56. $23{,}067 \times 6 =$

57. $99{,}090 \times 9 =$

58.
$$75 \times 25$$

59.
$$46 \times 39$$

60.
$$476 \times 52$$

61.
$$607 \times 44$$

62.
$$4{,}702 \times 73$$

63.
$$8{,}066 \times 45$$

Multiply.

64.
$$6{,}835 \times 47 =$$

65.
$$9{,}054 \times 93 =$$

66.
$$20{,}067 \times 82 =$$

67.
$$35{,}023 \times 51 =$$

68.
$$\begin{array}{r} 900 \\ \times\ 623 \\ \hline \end{array}$$

69.
$$\begin{array}{r} 5{,}476 \\ \times\ 213 \\ \hline \end{array}$$

70.
$$\begin{array}{r} 18{,}460 \\ \times\ 725 \\ \hline \end{array}$$

71.
$$\begin{array}{r} 854 \\ \times\ 472 \\ \hline \end{array}$$

72.
$$\begin{array}{r} 1{,}497 \\ \times\ 501 \\ \hline \end{array}$$

73.
$$521 \times 203 =$$

74.
$$110 \times 907 =$$

75.
$$3{,}021 \times 201 =$$

76.
$$784 \times 101 =$$

77.
$$\begin{array}{r} 400 \\ \times\ 203 \\ \hline \end{array}$$

78.
$$\begin{array}{r} 1{,}804 \\ \times\ 409 \\ \hline \end{array}$$

79.
$$\begin{array}{r} 78{,}009 \\ \times\ 104 \\ \hline \end{array}$$

80.
$$\begin{array}{r} 39{,}000 \\ \times\ 506 \\ \hline \end{array}$$

Below is a list of the problems in this review and the pages on which the skills are taught. If you missed any problems, turn to the pages listed and practice the skills. Then correct the problems you missed in the Unit Review.

Problems	Pages	Problems	Pages
1-8	85-87	38-47	101-102
9-16	89	48-53	105-106
17-22	91	54-57	107-108
23-24	95	58-63	110
25-32	96	64-67	111
33-37	97-98	68-72	112-113
		73–80	114-115

Dividing whole numbers is the opposite of multiplying them. Instead of finding a total, you split a total into equal groups. When 4 people split the cost of lunch, they divide the total cost equally by 4 to find the amount each person will pay.

In this unit, you will learn the division facts, and how to divide by one-digit, two-digit, and three-digit numbers. You will also learn about dividing with zeros.

Getting Ready

You should be familiar with the skills on this page and the next before you begin this unit. To check your answers, turn to page 194.

 You will need to round to the tens and hundreds places when you divide larger numbers.

Round each number to the nearest ten.

1. 14 ___**10**___ 2. 27 _____ 3. 45 _____ 4. 73 _____

Round each number to the nearest hundred.

5. 186 ___**200**___ 6. 550 _____ 7. 641 _____ 8. 397 _____

For review, see Unit 1, pages 20-21.

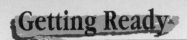

 To set up a subtraction problem, first line up the digits. Then solve the problem.

Line up the digits in each problem and solve.

9.
$$74 - 72 =$$
$$\begin{array}{r} 74 \\ -\ 72 \\ \hline 2 \end{array}$$

10.
$$468 - 354 =$$

11.
$$16 - 16 =$$

12.
$$599 - 490 =$$

For review, see Unit 3, pages 60-62.

 When subtracting from zero, borrow from the next column. You may need to borrow two or more times.

Solve.

13.
$$\begin{array}{r} {\scriptstyle 3\,10} \\ 24\!\!\!/0 \\ -\ 137 \\ \hline 103 \end{array}$$

14.
$$\begin{array}{r} 5,000 \\ -\ 1,657 \\ \hline \end{array}$$

15.
$$\begin{array}{r} 13,002 \\ -\ 7,937 \\ \hline \end{array}$$

16.
$$\begin{array}{r} 9,026 \\ -\ 8,900 \\ \hline \end{array}$$

For review, see Unit 3, pages 72-78.

 Knowing the multiplication facts will help you with division and with checking your answers.

Complete the following multiplication facts and multiplication problems.

17.
$$6 \times \boxed{5} = 30$$

18.
$$9 \times \boxed{} = 72$$

19.
$$3 \times \boxed{} = 21$$

20.
$$5 \times \boxed{} = 40$$

21.
$$32 \times 6 =$$
$$\begin{array}{r} 32 \\ \times\ 6 \\ \hline 192 \end{array}$$

22.
$$47 \times 69 =$$

23.
$$1,130 \times 25 =$$

24.
$$136 \times 259 =$$

25.
$$\boxed{} \times 7 = 56$$

26.
$$\boxed{} \times 4 = 12$$

27.
$$\boxed{} \times 9 = 81$$

28.
$$\boxed{} \times 6 = 48$$

29.
$$653 \times 45 =$$

30.
$$98 \times 37 =$$

31.
$$4,571 \times 126 =$$

32.
$$502 \times 312 =$$

For review, see Unit 4, pages 85-98.

Division Facts

To divide larger numbers, you should first know the basic division facts. You will find it helpful to know the following facts by heart.

Divide the following numbers to complete each row. Notice that the answers form a pattern.

1. $1\overline{)0}$ $1\overline{)1}$ $1\overline{)2}$ $1\overline{)3}$ $1\overline{)4}$ $1\overline{)5}$ $1\overline{)6}$ $1\overline{)7}$ $1\overline{)8}$ $1\overline{)9}$

2. $2\overline{)0}$ $2\overline{)2}$ $2\overline{)4}$ $2\overline{)6}$ $2\overline{)8}$ $2\overline{)10}$ $2\overline{)12}$ $2\overline{)14}$ $2\overline{)16}$ $2\overline{)18}$

3. $3\overline{)0}$ $3\overline{)3}$ $3\overline{)6}$ $3\overline{)9}$ $3\overline{)12}$ $3\overline{)15}$ $3\overline{)18}$ $3\overline{)21}$ $3\overline{)24}$ $3\overline{)27}$

4. $4\overline{)0}$ $4\overline{)4}$ $4\overline{)8}$ $4\overline{)12}$ $4\overline{)16}$ $4\overline{)20}$ $4\overline{)24}$ $4\overline{)28}$ $4\overline{)32}$ $4\overline{)36}$

5. $5\overline{)0}$ $5\overline{)5}$ $5\overline{)10}$ $5\overline{)15}$ $5\overline{)20}$ $5\overline{)25}$ $5\overline{)30}$ $5\overline{)35}$ $5\overline{)40}$ $5\overline{)45}$

6. $6\overline{)0}$ $6\overline{)6}$ $6\overline{)12}$ $6\overline{)18}$ $6\overline{)24}$ $6\overline{)30}$ $6\overline{)36}$ $6\overline{)42}$ $6\overline{)48}$ $6\overline{)54}$

7. $7\overline{)0}$ $7\overline{)7}$ $7\overline{)14}$ $7\overline{)21}$ $7\overline{)28}$ $7\overline{)35}$ $7\overline{)42}$ $7\overline{)49}$ $7\overline{)56}$ $7\overline{)63}$

8. $8\overline{)0}$ $8\overline{)8}$ $8\overline{)16}$ $8\overline{)24}$ $8\overline{)32}$ $8\overline{)40}$ $8\overline{)48}$ $8\overline{)56}$ $8\overline{)64}$ $8\overline{)72}$

9. $9\overline{)0}$ $9\overline{)9}$ $9\overline{)18}$ $9\overline{)27}$ $9\overline{)36}$ $9\overline{)45}$ $9\overline{)54}$ $9\overline{)63}$ $9\overline{)72}$ $9\overline{)81}$

Division Facts Practice

You can use the multiplication facts table to complete division facts since multiplication and division are opposite operations.

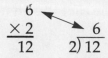

Use These Steps

Complete the division fact   20 ÷ 5

1. First, find the smaller number, 5, in the farthest row on the left.

2. Next, move to the right along that same row until you find the larger number, 20.

3. Then move to the top of the column to find the answer, 4.

÷	0	1	2	3	4	5	6	7	8	9
1	0	1	2	3	4	5	6	7	8	9
2	0	2	4	6	8	10	12	14	16	18
3	0	3	6	9	12	15	18	21	24	27
4	0	4	8	12	16	20	24	28	32	36
5	0	5	10	15	**20**	25	30	35	40	45
6	0	6	12	18	24	30	36	42	48	54
7	0	7	14	21	28	35	42	49	56	63
8	0	8	16	24	32	40	48	56	64	72
9	0	9	18	27	36	45	54	63	72	81

Use the table to complete the following division facts.

1. $2\overline{)8}$ (4)

2. $7\overline{)14}$

3. $6\overline{)24}$

4. $3\overline{)9}$

5. $6\overline{)30}$

6. $6\overline{)36}$

7. $8\overline{)16}$

8. $7\overline{)49}$

9. $8\overline{)56}$

10. $9\overline{)45}$

11. $7\overline{)63}$

12. $9\overline{)81}$

13. $6\overline{)48}$

14. $7\overline{)42}$

15. $3\overline{)24}$

16. $5\overline{)35}$

17. $4\overline{)12}$

18. $4\overline{)32}$

Division Facts Practice

Division problems can be written two ways.

$$\text{quotient} \longrightarrow \dfrac{4}{3)\,12} \quad \text{is the same as} \quad 12 \div 3 = 4$$

quotient $\longrightarrow$ 4
divisor $\longrightarrow$ 3)12
dividend

Complete the following division facts. You can use the table on page 124 if you need help remembering the facts.

1. $16 \div 8 = \boxed{2}$

2. $12 \div 4 = \boxed{}$

3. $0 \div 3 = \boxed{}$

4. $18 \div 9 = \boxed{}$

5. $20 \div \boxed{5} = 4$

6. $24 \div \boxed{} = 4$

7. $30 \div \boxed{} = 5$

8. $27 \div \boxed{} = 3$

9. $\boxed{} \div 7 = 5$

10. $\boxed{} \div 6 = 7$

11. $\boxed{} \div 9 = 5$

12. $\boxed{} \div 8 = 0$

13. $72 \div 8 = \boxed{}$

14. $81 \div \boxed{} = 9$

15. $64 \div 8 = \boxed{}$

16. $10 \div \boxed{} = 2$

17. $\boxed{} \div 2 = 7$

18. $20 \div \boxed{} = 5$

19. $\boxed{} \div 7 = 3$

20. $25 \div 5 = \boxed{}$

21. $27 \div \boxed{} = 3$

22. $30 \div 5 = \boxed{}$

23. $32 \div \boxed{} = 8$

24. $\boxed{} \div 6 = 6$

25. $\boxed{} \div 8 = 5$

26. $42 \div 6 = \boxed{}$

27. $45 \div \boxed{} = 9$

28. $54 \div \boxed{} = 9$

29. $56 \div \boxed{} = 7$

30. $\boxed{} \div 8 = 8$

31. $14 \div 7 = \boxed{}$

32. $63 \div 9 = \boxed{}$

Fill in the boxes with numbers that complete the division facts. There may be more than one set of numbers that makes a true statement.

33. $\boxed{6} \div \boxed{2} = 3$

34. $\boxed{} \div \boxed{} = 2$

35. $\boxed{} \div \boxed{} = 4$

36. $\boxed{} \div \boxed{} = 0$

37. $\boxed{} \div \boxed{} = 8$

38. $\boxed{} \div \boxed{} = 7$

39. $\boxed{} \div \boxed{} = 5$

40. $\boxed{} \div \boxed{} = 9$

41. $\boxed{} \div \boxed{} = 6$

Division as the Opposite of Multiplication

Dividing is the opposite of multiplying. This means that you can check the answer to a division problem by multiplying the answer by the number you divided by. The answer should be the same as the number you divided into.

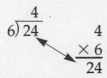

Use These Steps

Divide   32 ÷ 8

1. Set up the problem.

$$8\overline{)32}$$

2. Divide.

$$8\overline{)32}^{\,4}$$

3. Check by multiplying the answer, 4, by the number you divided by, 8. The answer should be the same as the number you divided into, 32.

Divide. Use multiplication to check your answers.

1.

$$4\overline{)20}^{\,5} \qquad \begin{array}{r} 5 \\ \times\,4 \\ \hline 20 \end{array}$$

2.

$$6\overline{)36}$$

3.

$$5\overline{)45}$$

4.

$$3\overline{)27}$$

5.

$$7\overline{)35}$$

6.

$$4\overline{)28}$$

7.

$$2\overline{)18}$$

8.

$$1\overline{)8}$$

9. $0 \div 5 =$

10. $42 \div 6 =$

11. $56 \div 8 =$

12. $32 \div 4 =$

13. $40 \div 8 =$

14. $49 \div 7 =$

15. $64 \div 8 =$

16. $72 \div 8 =$

Dividing by One-Digit Numbers

When you divide by a one-digit number, use the division facts. Be sure to put each answer in the correct column.

Use These Steps

Divide $3\overline{)249}$

1. Divide. Since you can't divide 2 by 3 evenly, divide 24 by 3. $24 \div 3 = 8$. Write the 8 above the 4.

$$\begin{array}{r} 8 \\ 3\overline{)249} \end{array}$$

⤒——— $24 \div 3 = 8$

2. Divide again. $9 \div 3 = 3$. Write the 3 above the 9.

$$\begin{array}{r} 83 \\ 3\overline{)249} \end{array}$$

3. Check by multiplying the answer, 83, by the number you divided by, 3.

$$\begin{array}{r} 83 \\ 3\overline{)249} \end{array}$$

$$\begin{array}{r} 83 \\ \times\ 3 \\ \hline 249 \end{array}$$

Divide. Use multiplication to check your answers.

1.
$$\begin{array}{r} 4 \\ 4\overline{)16} \end{array}$$
$$\begin{array}{r} 4 \\ \times\ 4 \\ \hline 16 \end{array}$$

2. $5\overline{)45}$

3. $7\overline{)63}$

4. $9\overline{)54}$

5. $5\overline{)105}$

6. $4\overline{)128}$

7. $2\overline{)144}$

8. $8\overline{)488}$

9. $6\overline{)126}$

10. $9\overline{)189}$

11. $3\overline{)213}$

12. $5\overline{)255}$

13. $186 \div 2 =$

14. $168 \div 4 =$

15. $189 \div 3 =$

16. $568 \div 8 =$

17. $637 \div 7 =$

18. $426 \div 6 =$

19. $305 \div 5 =$

20. $128 \div 2 =$

Dividing by One-Digit Numbers Using Long Division

Long division has several steps: divide, multiply, subtract, and bring down.

Divide $8\overline{)256}$

1. **Divide.** Since you can't divide 2 by 8 evenly, divide 25 by 8. $25 \div 8$ is not a basic fact. Use the closest fact. $24 \div 8 = 3$, so $25 \div 8 = 3$, plus an amount left over. Write 3 above the 5.

$$\begin{array}{r} 3 \\ 8\overline{)256} \end{array}$$

2. **Multiply.** $3 \times 8 = 24$. Subtract 24 from 25 to find the amount left over. The amount left must be less than 8. $25 - 24 = 1$. Bring down the next digit, 6.

$$\begin{array}{r} 3 \\ 8\overline{)256} \\ -24\downarrow \\ \hline 16 \end{array}$$

3. **Divide.** $16 \div 8 = 2$. Write 2 above the 6. Multiply. Subtract. Check the answer.

$$\begin{array}{r} 32 \\ 8\overline{)256} \\ -24 \\ \hline 16 \\ -16 \\ \hline 0 \end{array} \qquad \begin{array}{r} 32 \\ \times\ 8 \\ \hline 256 \end{array}$$

Divide.

1.
$$\begin{array}{r} 42 \\ 7\overline{)294} \\ -28 \\ \hline 14 \\ -14 \\ \hline 0 \end{array} \qquad \begin{array}{r} 42 \\ \times\ 7 \\ \hline 294 \end{array}$$

2. $6\overline{)384}$

3. $5\overline{)275}$

4. $4\overline{)184}$

5. $9\overline{)738}$

6. $3\overline{)282}$

7. $2\overline{)172}$

8. $8\overline{)432}$

9. $2\overline{)172}$

10. $5\overline{)475}$

11. $3\overline{)168}$

12. $6\overline{)288}$

13. $7\overline{)224}$

14. $6\overline{)318}$

15. $4\overline{)268}$

16. $8\overline{)416}$

Dividing by One-Digit Numbers with Remainders

When you divide, you will sometimes have an amount left over. This amount is called a *remainder*. Use the letter *R* to stand for remainder. The remainder is part of the answer.

Use These Steps

Divide $9\overline{)28}$

1. Divide. $28 \div 9$ is not a basic fact. Use the closest basic fact. $27 \div 9 = 3$, so $28 \div 9 = 3$, plus an amount left over. Write the 3 above the 8.

$$\begin{array}{r} 3 \\ 9\overline{)28} \end{array}$$

2. Multiply. $3 \times 9 = 27$. Subtract 27 from 28 to find the amount left over. The amount left over must be less than the number you divided by. $28 - 27 = 1$. The remainder is 1. Write R1 in the answer.

$$\begin{array}{r} 3 \text{ R1} \\ 9\overline{)28} \\ -27 \\ \hline 1 \end{array}$$

3. Check by multiplying the answer by the number you divided by. Then add the remainder. The answer should be the same as the number you divided into.

$$\begin{array}{r} 3 \text{ R1} \\ 9\overline{)28} \\ -27 \\ \hline 1 \end{array} \qquad \begin{array}{r} 3 \\ \times 9 \\ \hline 27 \\ + 1 \\ \hline 28 \end{array}$$

Divide. Use multiplication to check your answers.

1.
$$\begin{array}{r} 5 \text{ R2} \\ 5\overline{)27} \\ -25 \\ \hline 2 \end{array} \qquad \begin{array}{r} 5 \\ \times 5 \\ \hline 25 \\ + 2 \\ \hline 27 \end{array}$$

2. $3\overline{)10}$

3. $9\overline{)30}$

4. $7\overline{)59}$

5. $6\overline{)20}$

6. $8\overline{)44}$

7. $2\overline{)11}$

8. $4\overline{)34}$

9. $36 \div 7 =$

10. $43 \div 6 =$

11. $88 \div 9 =$

12. $15 \div 2 =$

13. $59 \div 8 =$

14. $33 \div 4 =$

15. $29 \div 5 =$

16. $17 \div 3 =$

Dividing by One-Digit Numbers with Remainders

When you divide into a larger number, you need to repeat the division steps two or more times. With each division step, you may have an amount left over.

Use These Steps

Divide $5\overline{)187}$

1. Divide. Since you can't divide 1 by 5 evenly, divide 18 by 5. $18 \div 5 = 3$, plus an amount left over. Write the 3 above the 8. Multiply. $3 \times 5 = 15$. Write the 15 under the 18 and subtract. The amount left is 3.

$$\begin{array}{r} 3 \\ 5\overline{)187} \\ -15 \\ \hline 3 \end{array}$$

2. Divide again by bringing down the next digit, 7. The 7 brought down beside the 3 makes 37. $37 \div 5 = 7$, plus an amount left over. Multiply. $7 \times 5 = 35$. Write the 35 under the 37 and subtract. $37 - 35 = 2$. The remainder is 2.

$$\begin{array}{r} 37 \text{ R2} \\ 5\overline{)187} \\ -15\downarrow \\ \hline 37 \\ -35 \\ \hline 2 \end{array}$$

3. Check your answer by multiplying. Add the remainder.

$$\begin{array}{r} 37 \\ \times\ 5 \\ \hline 185 \\ +\ 2 \\ \hline 187 \end{array}$$

Divide. Use multiplication to check your answers.

1.
$$\begin{array}{r} 29 \text{ R5} \\ 7\overline{)208} \\ -14 \\ \hline 68 \\ -63 \\ \hline 5 \end{array} \qquad \begin{array}{r} 29 \\ \times\ 7 \\ \hline 203 \\ +\ 5 \\ \hline 208 \end{array}$$

2. $5\overline{)192}$

3. $6\overline{)272}$

4. $9\overline{)424}$

5. $380 \div 6 =$

6. $395 \div 4 =$

7. $657 \div 8 =$

8. $508 \div 7 =$

9. $541 \div 7 =$

10. $178 \div 3 =$

11. $139 \div 2 =$

12. $566 \div 6 =$

Dividing by One-Digit Numbers with Remainders

Sometimes you will be able to divide into the first digit of a number.

Use These Steps

Divide $2\overline{)433}$

1. Divide. $4 \div 2 = 2$. Write the 2 above the 4. Multiply. Subtract. Bring down 3.

$$\begin{array}{r} 2 \\ 2\overline{)433} \\ -4\downarrow \\ \hline 03 \end{array}$$

2. Divide. $3 \div 2 = 1$, plus an amount left over. Write 1 in the answer. Multiply. Subtract. Bring down 3.

$$\begin{array}{r} 21 \\ 2\overline{)433} \\ -4 \\ \hline 03 \\ -2\downarrow \\ \hline 13 \end{array}$$

3. Divide again. $13 \div 2 = 6$, plus an amount left over. Multiply. Subtract. The remainder is 1.

$$\begin{array}{r} 216 \text{ R1} \\ 2\overline{)433} \\ -4 \\ \hline 03 \\ -2 \\ \hline 13 \\ -12 \\ \hline 1 \end{array} \qquad \begin{array}{r} 216 \\ \times\ 2 \\ \hline 432 \\ +\ 1 \\ \hline 433 \end{array}$$

Divide.

1.
$$\begin{array}{r} 214 \text{ R1} \\ 4\overline{)857} \\ -8 \\ \hline 05 \\ -4 \\ \hline 17 \\ -16 \\ \hline 1 \end{array} \qquad \begin{array}{r} 214 \\ \times\ 4 \\ \hline 856 \\ +\ 1 \\ \hline 857 \end{array}$$

2. $3\overline{)674}$

3. $2\overline{)479}$

4. $4\overline{)875}$

5. $9{,}486 \div 4 =$

6. $7{,}991 \div 3 =$

7. $6{,}789 \div 6 =$

8. $8{,}979 \div 8 =$

9. $6{,}687 \div 5 =$

10. $9{,}432 \div 7 =$

11. $3{,}527 \div 2 =$

12. $7{,}769 \div 6 =$

Real-Life Application   At Home

You can use division to help you split large amounts into smaller, equal groups.

Example   A bucket of chicken has 18 pieces. If you divide the pieces of chicken equally among 6 people, how many pieces does each person get?
To find equal amounts, divide.
$$18 \div 6 = 3$$
Each person will get 3 pieces of chicken.

Solve.

1. Bob, Tim, Alan, and Scott share a large house. They split their monthly expenses equally. Last month their electric bill was $144. How much did each of them pay?

2. Last month their water bill was $24. How much did each of them pay?

Answer _____

Answer _____

3. The rent on their house is $580 per month. How much rent does each of them pay every month?

4. The 4 roommates are thinking about asking their friend Tom to move in with them. If 5 people share the rent, how much will each person pay per month?

Answer _____

Answer _____

Mixed Review

Add, subtract, multiply, or divide.

1. $7\overline{)56}$

2. $\begin{array}{r} 5 \\ \times\, 6 \\ \hline \end{array}$

3. $8\overline{)72}$

4. $\begin{array}{r} 15 \\ -\ 7 \\ \hline \end{array}$

5. $6\overline{)42}$

6. $\begin{array}{r} 9 \\ +\, 8 \\ \hline \end{array}$

7. $24 \div 6 =$

8. $12 - 4 =$

9. $5\overline{)205}$

10. $\begin{array}{r} 32 \\ \times\, 4 \\ \hline \end{array}$

11. $4\overline{)288}$

12. $\begin{array}{r} 60 \\ +\, 9 \\ \hline \end{array}$

13. $203 \times 3 =$

14. $819 \div 9 =$

15. $355 - 5 =$

16. $147 \div 7 =$

17. $5\overline{)14}$

18. $3\overline{)23}$

19. $\begin{array}{r} 57 \\ +\, 9 \\ \hline \end{array}$

20. $4\overline{)30}$

21. $\begin{array}{r} 50 \\ \times\, 5 \\ \hline \end{array}$

22. $29 \div 8 =$

23. $52 \times 7 =$

24. $33 \div 4 =$

25. $20 - 3 =$

26. $\begin{array}{r} 330 \\ \times\, 6 \\ \hline \end{array}$

27. $2\overline{)196}$

28. $5\overline{)188}$

29. $\begin{array}{r} 670 \\ -\ 8 \\ \hline \end{array}$

30. $520 \div 8 =$

31. $174 \times 3 =$

32. $549 + 7 =$

33. $353 \div 4 =$

Dividing Larger Numbers

To divide larger numbers, follow the steps you learned on pages 129-131. You will need to divide two or more times.

Use These Steps

Divide $9\overline{)3,890}$

1. Divide. $38 \div 9 = 4$, plus an amount left over. Multiply. $4 \times 9 = 36$. Subtract. $38 - 36 = 2$. Bring down the next digit, 9.

$$\begin{array}{r} 4 \\ 9\overline{)3,890} \\ -36\downarrow \\ \hline 29 \end{array}$$

2. Divide. $29 \div 9 = 3$, plus an amount left over. Multiply. $3 \times 9 = 27$. Subtract. $29 - 27 = 2$. Bring down the 0.

$$\begin{array}{r} 43 \\ 9\overline{)3,890} \\ -36 \\ \hline 29 \\ -27\downarrow \\ \hline 20 \end{array}$$

3. Divide. $20 \div 9 = 2$, plus an amount left over. Multiply. $2 \times 9 = 18$. Subtract. $20 - 18 = 2$. Write R2 in the answer. Check your answer.

$$\begin{array}{r} 432 \text{ R2} \\ 9\overline{)3,890} \\ -36 \\ \hline 29 \\ -27 \\ \hline 20 \\ -18 \\ \hline 2 \end{array}$$

$$\begin{array}{r} 432 \\ \times 9 \\ \hline 3,888 \\ + 2 \\ \hline 3,890 \end{array}$$

Divide. Use multiplication to check your answers.

1.
$$\begin{array}{r} 438 \text{ R5} \\ 8\overline{)3,509} \\ -32 \\ \hline 30 \\ -24 \\ \hline 69 \\ -64 \\ \hline 5 \end{array}$$
$$\begin{array}{r} 438 \\ \times 8 \\ \hline 3\ 504 \\ + 5 \\ \hline 3,509 \end{array}$$

2. $6\overline{)1,614}$

3. $5\overline{)2,137}$

4. $3\overline{)1,934}$

5. $2,935 \div 4 =$

6. $5,828 \div 8 =$

7. $1,458 \div 6 =$

8. $1,843 \div 5 =$

9. $1,779 \div 2 =$

10. $3,286 \div 7 =$

11. $2,533 \div 4 =$

12. $3,818 \div 9 =$

Dividing Larger Numbers

Remember, there are four parts to each division step: divide, multiply, subtract, and bring down the next digit.

Use These Steps

Divide   1,868 ÷ 6

1. Set up the problem. Divide. 18 ÷ 6 = 3. Multiply. Subtract. 18 − 18 = 0. Put the 0 under the 8. Bring down the 6.

```
     3
6)1,868
 −18↓
   06
```

2. Divide. 6 ÷ 6 = 1. Multiply. Subtract. Bring down the 8.

```
    31
6)1,868
 −18
   06
  − 6↓
    08
```

3. Divide. 8 ÷ 6 = 1, plus an amount left over. Multiply. Subtract. Write the remainder in the answer. Check your answer.

```
   311 R2
6)1,868          311
 −18           ×   6
   06          1,866
  − 6          +   2
    08         1,868
   − 6
     2
```

Divide. Use multiplication to check your answers.

1.
```
     361 R8
9)3,257        361
 −27         ×   9
  55         3,249
 −54         +   8
  17         3,257
 − 9
   8
```

2. 8)6,888

3. 7)42,859

4. 6)57,486

5. 18,942 ÷ 3 =

6. 26,476 ÷ 5 =

7. 67,589 ÷ 4 =

8. 42,953 ÷ 6 =

9. 18,431 ÷ 2 =

10. 82,136 ÷ 9 =

Dividing into Zeros

When you divide a number with one or more zeros in the dividend, use the same steps as with other division problems: divide, multiply, and subtract. Bring down the next digit, even if it is a zero.

Use These Steps

Divide 5)1,105

1. Divide. 11 ÷ 5 = 2, plus an amount left over. Multiply, subtract, and bring down the 0.

```
      2
  5)1,105
   -10↓
     10
```

2. Divide. 10 ÷ 5 = 2. Multiply and subtract. Be sure to write the 0. Bring down the 5.

```
     22
  5)1,105
   -10
     10
    -10↓
      05
```

3. Divide. 5 ÷ 5 = 1. Multiply and subtract. There is no remainder. Check your answer.

```
    221              221
 5)1,105           ×   5
  -10              1,105
    10
   -10
     05
    - 5
      0
```

Divide. Use multiplication to check your answers.

1.
```
     15
 7) 105        15
  - 7         × 7
    35        105
  - 35
     0
```

2. 8) 209

3. 6) 504

4. 5) 406

5.
6,008 ÷ 8 =
```
    751
 8) 6,008      751
  - 56        ×   8
    40        6,008
  - 40
    08
   - 8
     0
```

6. 2,008 ÷ 7 =

7. 5,003 ÷ 6 =

8. 13,007 ÷ 8 =

9. 16,076 ÷ 3 =

10. 10,006 ÷ 6 =

Dividing into Zeros

The number you are dividing into may have 1 or more zeros. Use the same steps as in other division problems: divide, multiply, subtract, and bring down. Bring down the next digit, even if it is a zero. When dividing into zero, write 0 in the answer.

Use These Steps

Divide $9\overline{)8{,}100}$

1. Divide. $81 \div 9 = 9$. Multiply and subtract. Bring down the 0.

$$
\begin{array}{r}
9 \\
9\overline{)8{,}100} \\
-81\downarrow \\
\hline
00
\end{array}
$$

2. Divide. $0 \div 9 = 0$. Multiply and subtract. Write 0 in the answer. Bring down the next 0.

$$
\begin{array}{r}
90 \\
9\overline{)8{,}100} \\
-81 \\
\hline
00 \\
-0\downarrow \\
\hline
00
\end{array}
$$

3. Divide. $0 \div 9 = 0$. Multiply and subtract. Write 0 in the answer. Check your answer.

$$
\begin{array}{r}
900 \\
9\overline{)8{,}100} \\
-81 \\
\hline
00 \\
-0 \\
\hline
00 \\
-0 \\
\hline
0
\end{array}
\qquad
\begin{array}{r}
900 \\
\times\ \ 9 \\
\hline
8{,}100
\end{array}
$$

Divide.

1.
$$
\begin{array}{r}
400 \\
4\overline{)1{,}600} \\
-16 \\
\hline
00 \\
-0 \\
\hline
00 \\
-0 \\
\hline
0
\end{array}
\qquad
\begin{array}{r}
400 \\
\times\ \ 4 \\
\hline
1{,}600
\end{array}
$$

2. $5\overline{)3{,}000}$

3. $6\overline{)2{,}400}$

4. $2\overline{)4{,}000}$

5. $64{,}000 \div 8 =$

6. $42{,}000 \div 7 =$

7. $18{,}000 \div 3 =$

8. $36{,}000 \div 9 =$

9. $48{,}000 \div 6 =$

10. $81{,}000 \div 9 =$

11. $42{,}000 \div 6 =$

12. $56{,}000 \div 7 =$

Problem Solving: Using Rounding and Estimating

When you are dividing one large payment into several small payments, you can use rounding and estimating.

> **Example** Audrey is buying a stereo. The stereo she wants costs $298. Instead of paying for it all at once, she wants to take 6 months to pay for the stereo. About how much will Audrey have to pay each month?

▶ **Step 1.** Round the cost of the stereo to the nearest hundred.
$298 rounds to $300.

▶ **Step 2.** Divide to estimate the monthly payments.

$$
\begin{array}{r}
50 \\
6)\overline{300} \\
-\underline{30} \\
00
\end{array}
$$

Audrey will pay about $50 each month for 6 months.

Solve by rounding to the nearest hundred. Then divide.

1. Tina bought a bicycle for $209. If she pays for the bicycle in 5 monthly payments, about how much will she pay each month?

2. Teresa bought an antique table for $378. She will pay for it in 8 monthly payments. About how much will she pay each month?

Answer _____

Answer _____

3. Buddy is planning to buy a couch for $599. He wants to make 5 monthly payments. About how much will he pay each month?

4. If Buddy decides to make 8 monthly payments for the couch, about how much will he pay each month?

Answer _____

Answer _____

5. Joe wants to buy a used car that costs $4,819. If he splits the cost into 2 payments, about how much will each payment be?

Answer _____

6. If Joe pays for the car in 6 months, about how much will he pay each month?

Answer _____

7. Joe has decided that he can afford to pay for the car in 4 monthly payments. About how much will he pay each month?

Answer _____

8. Brian is buying a truck for $8,215. If he splits the cost into 8 payments, about how much will each payment be?

Answer _____

9. Ellen has to buy car insurance. The insurance costs $619 for one year. If Ellen decides to split the cost into 2 payments, about how much will each payment be?

Answer _____

10. If Ellen decides to pay for her insurance by the month, about how much will she pay each month? (Hint: There are 12 months in a year.)

Answer _____

11. Mary and Louisa are sharing an apartment. The rent is $445 each month. If they split the rent, about how much does each of them pay each month?

Answer _____

12. Louisa wants a cat. She can have a cat if she pays a pet deposit of $175. If Louisa pays the deposit in 4 payments, about how much will each payment be?

Answer _____

Dividing by Two-Digit Numbers

When you divide by a two-digit number, you may need to estimate.

Use These Steps

Divide $25\overline{)95}$

1. Divide 95 by 25 by estimating how many times 2 goes into 9. $9 \div 2$ is about 4. Write the 4 above the 5. Multiply. $4 \times 25 = 100$. 100 is larger than 95, so 4 is too large.

$$25\overline{)95}^{\,4}$$

2. Try a smaller number, 3. $25 \times 3 = 75$. 75 is smaller than 95. Subtract. $95 - 75 = 20$. The answer is 3, plus a remainder of 20.

$$\begin{array}{r} 3\text{ R}20 \\ 25\overline{)95} \\ -75 \\ \hline 20 \end{array}$$

3. Check your answer.

$$\begin{array}{r} 25 \\ \times\ 3 \\ \hline 75 \\ +\ 20 \\ \hline 95 \end{array}$$

Divide. Use multiplication to check your answers.

1.
$$\begin{array}{r} 3\text{ R}2 \\ 27\overline{)83} \\ -81 \\ \hline 2 \end{array} \qquad \begin{array}{r} 27 \\ \times\ 3 \\ \hline 81 \\ +\ 2 \\ \hline 83 \end{array}$$

2. $12\overline{)50}$

3. $31\overline{)93}$

4. $45\overline{)90}$

5. $24\overline{)360}$

6. $33\overline{)530}$

7. $56\overline{)677}$

8. $62\overline{)809}$

9. $566 \div 19 =$

10. $996 \div 83 =$

11. $702 \div 48 =$

12. $800 \div 61 =$

Dividing by Two-Digit Numbers

When you divide by a two digit number, first estimate, then multiply. If the answer is too large, try dividing by a smaller number.

Use These Steps

Divide $34)\overline{2,096}$

1. Divide. Since you can't divide 20 by 34 evenly, divide 209 by 34. Estimate how many times 3 goes into 20. 20 ÷ 3 is about 6. Try 6. Put the 6 above the 9. Multiply. 6 × 34 = 204. Subtract. Bring down the next digit.

2. Divide. 5 ÷ 3 is about 1. Multiply. Subtract. The answer is 61, plus a remainder of 22.

3. Check your answer.

$$
\begin{array}{r}
6 \\
34)\overline{2,096} \\
-\,2\,04\downarrow \\
\hline
56
\end{array}
$$

$$
\begin{array}{r}
61\ R22 \\
34)\overline{2,096} \\
-\,2\,04 \\
\hline
56 \\
-\,34 \\
\hline
22
\end{array}
$$

$$
\begin{array}{r}
61 \\
\times\ 34 \\
\hline
244 \\
+\,1\,83 \\
\hline
2,074 \\
+\quad 22 \\
\hline
2,096
\end{array}
$$

Divide. Use multiplication to check your answers.

1.
$$
\begin{array}{r}
3\ R24 \\
32)\overline{120} \\
-\,96 \\
\hline
24
\end{array}
\qquad
\begin{array}{r}
32 \\
\times\ 3 \\
\hline
96 \\
+\,24 \\
\hline
120
\end{array}
$$

2. $22)\overline{162}$

3. $46)\overline{351}$

4. $54)\overline{327}$

5. $96)\overline{4,320}$

6. $56)\overline{1,650}$

7. $36)\overline{1,900}$

8. $93)\overline{8,026}$

9. $6,550 \div 75 =$

10. $3,000 \div 42 =$

11. $7,452 \div 96 =$

Dividing by Two-Digit Numbers

When you divide by two-digit numbers, line up the digits in the correct columns for the subtraction step. If you round the number you are dividing by, you will get an estimate.

Use These Steps

Divide $59\overline{)13,865}$

1. Divide. 59 rounds to 60. Use 6 to estimate each answer. Multiply. Subtract, and bring down.

2. Check your answer.

```
        2                23               235
59 )13,865        59 )13,865        59 )13,865
   − 11 8↓            − 11 8             − 11 8
     2 06              2 06               2 06
                      − 1 77↓            − 1 77
                        295                295
                                         − 295
                                             0
```

```
    235
  ×  59
  2 115
 + 11 75
 13,865
```

Divide. Use multiplication to check your answers.

1.
```
      169  R18
32 )5,426          169
   − 3 2          × 32
    2 22           338
   − 1 92         + 5 07
     306          5 408
   − 288         +   18
      18          5,426
```

2. $51\overline{)4,896}$

3. $65\overline{)3,371}$

4. $29\overline{)3,326}$

5. $47\overline{)52,687}$

6. $38\overline{)28,559}$

7. $26\overline{)14,726}$

8. $63\overline{)73,017}$

9. $26,000 \div 32 =$

10. $65,200 \div 58 =$

11. $99,424 \div 52 =$

Dividing with Zeros

In a division problem, the number you divide by sometimes ends in zero. Be sure to include the zero when you multiply and subtract.

Use These Steps

Divide $80\overline{)44{,}083}$

1. Divide, multiply, subtract, and bring down.

```
      5              55             551 R3
80)44,083      80)44,083      80)44,083
  -40 0          -40 0          -40 0
   4 08           4 08           4 08
                 -4 00          -4 00
                    83             83
                                 - 80
                                    3
```

2. Check your answer.

```
      551
    ×  80
   44,080
   +    3
   44,083
```

Divide. Use multiplication to check your answers.

1.
```
      35            35
10)350           × 10
  -30            350
   50
  -50
    0
```

2. $30\overline{)930}$

3. $50\overline{)4{,}225}$

4. $20\overline{)1{,}182}$

5. $60\overline{)3{,}900}$

6. $30\overline{)54{,}630}$

7. $10\overline{)29{,}340}$

8. $80\overline{)79{,}040}$

9. $25{,}007 \div 40 =$

10. $95{,}550 \div 50 =$

11. $86{,}310 \div 10 =$

12. $17{,}040 \div 20 =$

13. $41{,}940 \div 90 =$

14. $91{,}876 \div 70 =$

Mixed Review

Add, subtract, multiply, or divide.

1.
$$\begin{array}{r} 15 \\ + 67 \\ \hline \end{array}$$

2.
$$\begin{array}{r} 48 \\ - 39 \\ \hline \end{array}$$

3.
$$\begin{array}{r} 56 \\ + 42 \\ \hline \end{array}$$

4. $31\overline{)217}$

5.
$$\begin{array}{r} 61 \\ \times 43 \\ \hline \end{array}$$

6. $25\overline{)300}$

7.
$$\begin{array}{r} 50 \\ - 37 \\ \hline \end{array}$$

8. $59\overline{)178}$

9.
$$\begin{array}{r} 24 \\ + 76 \\ \hline \end{array}$$

10. $86\overline{)774}$

11.
$$\begin{array}{r} 99 \\ \times 30 \\ \hline \end{array}$$

12. $980 \div 20 =$

13. $70 \times 70 =$

14. $80 - 10 =$

15. $60 + 70 =$

16.
$$\begin{array}{r} 362 \\ \times 45 \\ \hline \end{array}$$

17. $50\overline{)9,132}$

18.
$$\begin{array}{r} 403 \\ - 99 \\ \hline \end{array}$$

19. $95\overline{)2,470}$

20. $15\overline{)8,439}$

21.
$$\begin{array}{r} 478 \\ \times 96 \\ \hline \end{array}$$

22.
$$\begin{array}{r} 500 \\ - 452 \\ \hline \end{array}$$

23. $40\overline{)3,440}$

Add, subtract, multiply, or divide.

24.
$$2{,}679 + 43 =$$

25.
$$2{,}359 \div 31 =$$

26.
$$4{,}320 \times 90 =$$

27.
$$41\overline{)29{,}563}$$

28.
$$\begin{array}{r} 5{,}432 \\ \times\ \ \ 126 \\ \hline \end{array}$$

29.
$$\begin{array}{r} 4{,}792 \\ -\ 1{,}849 \\ \hline \end{array}$$

30.
$$\begin{array}{r} 6{,}540 \\ +\ 9{,}873 \\ \hline \end{array}$$

31.
$$10\overline{)83{,}270}$$

32.
$$46\overline{)5{,}402}$$

33.
$$\begin{array}{r} 8{,}729 \\ \times\ \ \ 557 \\ \hline \end{array}$$

34.
$$\begin{array}{r} 9{,}000 \\ -\ 7{,}430 \\ \hline \end{array}$$

35.
$$5{,}486 + 297 + 54 =$$

36.
$$27{,}800 - 18{,}922 =$$

37.
$$99{,}800 \div 42 =$$

38.
$$2{,}000 \times 409 =$$

39.
$$80\overline{)50{,}320}$$

40.
$$35\overline{)26{,}430}$$

41.
$$\begin{array}{r} 25{,}040 \\ \times\ \ \ \ \ 37 \\ \hline \end{array}$$

42.
$$\begin{array}{r} 70{,}000 \\ -\ 3{,}428 \\ \hline \end{array}$$

Real-Life Application

When you want to find the best buy at the grocery store, you compare prices. To do this, you need to know the unit cost of items. For example, if soup is on sale at 3 cans for 99 cents, the unit cost is 99 ÷ 3, or 33 cents per can.

Example   Mealtime soup is on sale at 2 cans for 52 cents. If 3 cans of Valley Fresh soup sell for 99 cents, which soup costs more: Mealtime or Valley Fresh?

Divide to find the unit cost. Then compare.

Valley Fresh	Mealtime
33 cents	26 cents
3)99	2)52

33 cents > 26 cents

Valley Fresh soup costs more than Mealtime soup.

Solve.

1. Bunty paper towels cost 49 cents per roll. Diva paper towels are on sale at 2 for 94 cents. Which brand costs more: Bunty or Diva?

2. Bow Wow Chow dog food is on sale at 4 cans for 80 cents. Barker's dog food costs 85 cents for 5 cans. Which dog food costs more: Bow Wow Chow or Barker's?

Answer _____

Answer _____

3. An 8-ounce package of Curly's noodles costs 63 cents. A 10-ounce package of Italia noodles costs 90 cents. Which costs more: Curly's or Italia?

4. You can buy onions for 20 cents a pound, or you can buy a 3-pound bag for 57 cents. How much money will you save if you buy the 3-pound bag?

Answer _____

Answer _____

Dividing by Three-Digit Numbers

When you divide by a three-digit number, first estimate. Then multiply to see if your estimate is correct.

Use These Steps

Divide $341\overline{)9{,}930}$

1. Divide 993 by 341 by estimating how many times 3 goes into 9. $9 \div 3 = 3$. Multiply. $341 \times 3 = 1{,}023$. Try a smaller number. $341 \times 2 = 682$. Subtract. Bring down.

2. Divide 31 by 3 to estimate. $31 \div 3 = 10$. The number 10 will not fit in the answer. Try 9. $341 \times 9 = 3{,}069$. Subtract. Check your answer.

```
        2
 341)9,930
   - 6 82↓
     3 110
```

```
       29 R41
 341)9,930        341
   - 6 82       ×  29
     3 110      3 069
   - 3 069      + 6 82
        41      9,889
              +    41
                9,930
```

Divide. Use multiplication to check your answers.

1.
```
         13
 529)6,877        529
   - 5 29       ×  13
     1 587      1 587
   - 1 587      + 5 29
         0      6,877
```

2. $692\overline{)8{,}304}$

3. $478\overline{)6{,}697}$

4. $861\overline{)9{,}614}$

5. $525\overline{)9{,}450}$

6. $117\overline{)6{,}669}$

7. $9{,}297 \div 101 =$

8. $5{,}916 \div 348 =$

9. $6{,}488 \div 926 =$

Dividing by Three-Digit Numbers

When you subtract and bring down, be sure to keep the digits lined up in the correct columns.

Use These Steps

Divide $601 \overline{)369,615}$

1. Divide, multiply, subtract, and bring down.

```
        6                  61                 615
601) 369,615       601) 369,615       601) 369,615
   - 360 6↓           - 360 6 |          - 360 6
      9 01               9 01               9 01
                      -  6 01↓            -  6 01
                         3 005              3 005
                                         -  3 005
                                                0
```

2. Check your answer.

```
      601
    × 615
    3 005
    6 01
+ 360 6
  369,615
```

Divide. Use multiplication to check your answers.

1.
```
         318
915) 290,970            318
  - 274 5             × 915
    16 47             1 590
   -  9 15            3 18
     7 320          + 286 2
   -  7 320          290,970
         0
```

2. $653 \overline{)594,913}$

3. $491 \overline{)33,388}$

4. $97,498 \div 402 =$

5. $118,692 \div 756 =$

6. $94,552 \div 109 =$

7. $324 \overline{)264,859}$

8. $502 \overline{)59,514}$

9. $369 \overline{)176,751}$

Dividing with Zeros in the Answer

When you subtract and bring down, you sometimes get a number that is smaller than the number you are dividing into. When this happens, write a zero in the answer and bring down the next digit.

Use These Steps

Divide $5\overline{)3,010}$

1. Divide. Multiply. Subtract. Bring down.

$$
\begin{array}{r}
6 \\
5\overline{)3,010} \\
-3\,0\downarrow \\
\hline
01
\end{array}
$$

2. Five is larger than 1, so you can't divide. Put a zero in the answer, and bring down the next number.

$$
\begin{array}{r}
60 \\
5\overline{)3,010} \\
-3\,0\downarrow \\
\hline
010
\end{array}
$$

3. Divide. Multiply. Subtract. Check your answer.

$$
\begin{array}{r}
602 \\
5\overline{)3,010} \\
-3\,0 \\
\hline
010 \\
-10 \\
\hline
0
\end{array}
\qquad
\begin{array}{r}
602 \\
\times\quad 5 \\
\hline
3,010
\end{array}
$$

Divide. Use multiplication to check your answers.

1.
$$
\begin{array}{r}
203 \\
75\overline{)15,225} \\
-15\,0 \\
\hline
225 \\
-225 \\
\hline
0
\end{array}
\qquad
\begin{array}{r}
203 \\
\times\quad 75 \\
\hline
1\,015 \\
+14\,21 \\
\hline
15,225
\end{array}
$$

2. $7\overline{)7,014}$

3. $3\overline{)1,518}$

4. $28,210 \div 35 =$

5. $27,336 \div 68 =$

6. $77,125 \div 96 =$

7. $418\overline{)211,511}$

8. $523\overline{)53,869}$

9. $290\overline{)261,581}$

Dividing with Zeros in the Answer

When you divide, use zero as you would use any other digit.

Divide. Use multiplication to check your answers.

1.
```
        306 R2
28) 8,570
  − 8 40
    170
  − 168
      2
```
```
       306
     ×  28
     2 448
   + 61 2
     8 568
   +     2
     8,570
```

2. $9\overline{)54,023}$

3. $3\overline{)15,272}$

4. $76\overline{)61,409}$

5. $91\overline{)456,820}$

6. $882\overline{)793,814}$

7. $375,769 \div 501 =$

8. $243,405 \div 405 =$

9. $70,200 \div 702 =$

Real-Life Application On the Job

Elaine is a route manager for the city newspaper. She counts the number of newspapers each newsstand needs, delivers the bundles of papers, and collects the unsold papers and money from each stand.

Example On Monday, Elaine delivered 10,125 papers to 135 newsstands. Each newsstand gets the same number of newspapers. How many newspapers did Elaine deliver to each stand?

$$
\begin{array}{r}
75 \\
135)\overline{10{,}125} \\
-\ 9\ 45 \\
\hline
675 \\
-\ 675 \\
\hline
0
\end{array}
$$

Elaine delivered 75 newspapers to each stand.

Solve.

1. On Tuesday, Elaine delivered 15,600 papers. Each stand received 100 papers. How many stands did she deliver papers to?

2. Elaine delivered the same number of papers on Wednesday, Thursday, and Friday. If she delivered 58,995 papers in all, how many did she deliver each day?

Answer _____

Answer _____

3. On Sunday, Elaine delivered 20,150 papers. There are 25 papers to a bundle. How many bundles of papers did Elaine deliver?

4. Eduardo's newsstand took in a total of $300 for the Sunday paper. If Eduardo sold 150 papers, how much did he charge for each paper?

Answer _____

Answer _____

Unit 5 Review

Divide. Use multiplication to check your answers.

1. $6\overline{)18}$　　2. $8\overline{)24}$　　3. $9\overline{)72}$　　4. $5\overline{)45}$　　5. $7\overline{)42}$

6. $10 \div 2 =$　　7. $15 \div 3 =$　　8. $25 \div 5 =$　　9. $32 \div 8 =$　　10. $0 \div 9 =$

11. $2\overline{)24}$　　12. $3\overline{)396}$　　13. $5\overline{)555}$　　14. $9\overline{)369}$　　15. $6\overline{)426}$

16. $693 \div 3 =$　　17. $482 \div 2 =$　　18. $728 \div 8 =$

19. $5\overline{)145}$　　20. $6\overline{)324}$　　21. $8\overline{)1,864}$　　22. $7\overline{)377}$

23. $4\overline{)159}$　　24. $9\overline{)3,229}$　　25. $2\overline{)11,159}$

26. $4\overline{)500}$　　27. $7\overline{)3,024}$　　28. $75,705 \div 5$　　29. $10,023 \div 3$

Divide. Use multiplication to check your answers.

30.
$$14\overline{)28}$$

31.
$$22\overline{)66}$$

32.
$$32\overline{)96}$$

33.
$$42\overline{)84}$$

34.
$$54\overline{)486}$$

35.
$$87\overline{)2,267}$$

36.
$$65\overline{)5,050}$$

37.
$$91\overline{)19,201}$$

38.
$3,029 \div 42 =$

39.
$26,098 \div 59 =$

40.
$80,582 \div 86 =$

41.
$$60\overline{)3,950}$$

42.
$$10\overline{)2,165}$$

43.
$$30\overline{)9,360}$$

44.
$$50\overline{)16,450}$$

45.
$32,067 \div 90 =$

46.
$82,674 \div 70 =$

47.
$95,020 \div 20 =$

Divide. Use multiplication to check your answers.

48.
$326\overline{)815}$

49.
$270\overline{)339}$

50.
$792\overline{)862}$

51.
$530\overline{)862}$

52.
$521\overline{)3,824}$

53.
$862\overline{)1,588}$

54.
$711\overline{)25,622}$

55.
$451\overline{)23,001}$

56.
$3\overline{)32,983}$

57.
$412\overline{)8,360}$

58.
$330\overline{)16,500}$

59.
$54,120 \div 902 =$

60.
$95,000 \div 500 =$

61.
$50,000 \div 200 =$

Below is a list of the problems in this review and the pages on which the skills are taught. If you missed any problems, turn to the pages listed and practice the skills. Then correct the problems you missed in the Unit Review.

Problems	Pages	Problems	Pages
1-10	123-125	30-40	140-142
11-18	127	41-47	143
19-25	128-135	48-55	147-148
26-29	136-137	56-61	149-150

You have studied four operations with whole numbers — addition, subtraction, multiplication, and division. You have applied these skills to real-life problems, and you have learned techniques for solving word problems.

In this unit, you will study how to choose the correct operation needed to solve problems. You will also learn how to solve problems that require you to use more than one operation to find the answers.

Getting Ready

You should be familiar with the skills on this page and the next before you begin this unit. To check your answers, turn to page 201.

 To add whole numbers, be sure the digits are lined up. Start by adding the digits in the ones column. You may need to rename by carrying to the next column.

Add.

1.	2.	3.	4.	5.
15 + 32 **47**	25 + 47	162 + 89	103 + 47	386 + 499

6.	7.	8.	9.	10.
1,487 + 613	5,429 + 1,876	2,032 + 8,961	15,300 + 25,800	70,000 + 42,037

 To subtract whole numbers, be sure the digits are lined up. Start by subtracting the digits in the ones column. You may need to rename by borrowing from the next column.

Subtract.

11.	12.	13.	14.	15.
96	84	73	247	347
− 43	− 74	− 26	− 89	− 268
53				

16.	17.	18.	19.	20.
3,420	5,809	9,030	10,500	40,000
− 932	− 2,532	− 4,672	− 4,670	− 29,641

For review, see Unit 3.

 To multiply whole numbers, line up the digits. Start by multiplying by the ones digit. You may need to rename by carrying to the next column.

Multiply.

21.	22.	23.	24.	25.
32	20	41	89	90
× 4	× 30	× 36	× 3	× 36
128				

26.	27.	28.	29.	30.
138	507	1,406	420	2,065
× 30	× 53	× 198	× 307	× 100

For review, see Unit 4.

 To divide, set up the problem. Start by dividing into the digits on the left. You may need to bring down one or more digits.

Divide.

31.
```
      51
  5) 255
   − 25
      05
    −  5
       0
```

32. 7) 432

33. 9) 936

34. 62) 4,340

35. 41) 3,772

For review, see Unit 5.

Choose an Operation: Being a Consumer

Cholesterol is a fatty substance found in many foods. Many doctors say that to prevent heart disease, people should lower their cholesterol levels. One way to do this is to eat foods that are low in cholesterol. The chart shows the cholesterol content of some popular foods.

Example Gail is trying to lower her cholesterol level. What will be the difference in milligrams of cholesterol if she drinks 1 cup of skim milk instead of 1 cup of whole milk for breakfast?

▶ **Step 1.** To find the difference, subtract.

▶ **Step 2.** Use the chart to find the amounts of cholesterol.

1 cup of skim milk = 5 milligrams
1 cup of whole milk = 34 milligrams

▶ **Step 3.** Subtract.

$$34 - 5 = 29$$

Cholesterol Content in Milligrams (mgs) of Some Common Foods

	serving	cholesterol
skim milk	1 cup	5 mgs
cottage cheese	$\frac{1}{2}$ cup	7
ice cream	$\frac{1}{2}$ cup	28
cheddar cheese	1 ounce	28
whole milk	1 cup	34
butter	1 tablespoon	35
chicken	3 ounces	67
beef	3 ounces	75
egg	1	250

Gail will take in 29 fewer milligrams of cholesterol if she drinks skim milk.

Solve.

1. Gail had 1 cup of skim milk and 1 egg for breakfast. How many total milligrams of cholesterol were in her breakfast?

Answer_____

2. If Gail eats 2 servings of beef for dinner tomorrow, how many milligrams of cholesterol will she take in?

Answer_____

3. Gail ate 2 servings of chicken for dinner. Which expression would you use to find how many milligrams of cholesterol there are in 2 servings of chicken?

   a. 2×6 Solve for the answer.
   b. 67×2
   c. $67 + 2$
   d. $67 - 2$

Answer_____

4. Circle the expression you would use to find how many more milligrams of cholesterol there are in 3 ounces of beef than in 3 ounces of chicken.

   a. $67 + 3$ Solve for the answer.
   b. $67 \div 3$
   c. $75 - 67$
   d. 75×67

Answer_____

Choose an Operation: Using Measurement

To change from one unit of measurement to another, remember that when you change a small unit to a large unit, you divide. When you change a large unit to a small unit, you multiply.

12 inches	=	1 foot
3 feet	=	1 yard
1,760 yards	=	1 mile
5,280 feet	=	1 mile

Example Maude bought 48 inches of ribbon to make holiday bows. How many feet of ribbon did she buy?

▶ **Step 1.** You are changing inches to feet. To change a small unit to a large unit, divide.

▶ **Step 2.** Use the chart to find out how many inches there are in one foot.

$$12 \text{ inches} = 1 \text{ foot}$$

▶ **Step 3.** Divide.

$$48 \div 12 = 4 \text{ feet}$$

Solve.

1. Helena is 6 feet tall. How many inches tall is she?

   Answer_____

2. Michiko rode her bicycle 2 miles to the store. How many yards did she ride?

   Answer_____

3. Alberto bought some boards to use to repair his porch. Each board is 9 feet long. Circle the expression you would use to find how long each board is in inches.

   a. 9×3 Solve for the answer.
   b. $9 \div 3$
   c. 9×12
   d. $9 \div 12$

   Answer_____

4. Circle the expression you would use to find how long each of Alberto's boards is in yards.

   a. 9×3 Solve for the answer.
   b. $9 \div 3$
   c. $12 \div 3$
   d. 12×3

   Answer_____

Choose an Operation: Using a Bar Graph

The total monthly rainfall for a ten-year period in Petersburg is shown in the bar graph. The numbers in the column on the left show inches of rain. Each bar shows the amount of rainfall for that month.

Example   What was the total rainfall for April and May in Petersburg?

▶ **Step 1.**   To find the total, add.

▶ **Step 2.**   Use the graph to find the rainfall amounts.
They are shown by the height of each bar.

April = 10 inches / May = 30 inches

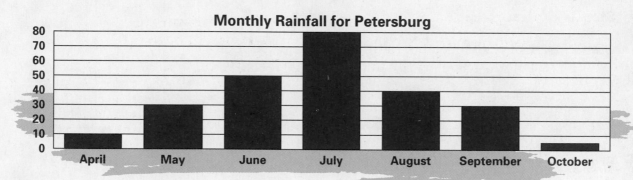

Monthly Rainfall for Petersburg

▶ **Step 3.**   Add.

10 + 30 = 40 inches

In April and May, 40 inches of rain fell in Petersburg.

Solve.

1. How much more rain fell in July than in June?

   Answer_____

2. The rainfall for August was exactly half the rainfall for which month?

   Answer_____

3. Circle the expression you would use to find how much more rain fell in April than in October.

   a. 10 + 10   Solve for the answer.
   b. 10 ÷ 5
   c. 10 − 0
   d. 10 − 5

   Answer_____

4. Circle the expression you would use to find how much total rain fell in July and August.

   a. 80 + 40   Solve for the answer.
   b. 40 × 80
   c. 80 − 40
   d. 80 ÷ 40

   Answer_____

You may need to use two operations to solve some problems.
Look at the city map. Each block is a square that is 500 feet long on each
side. Each intersection is marked with a letter. The shaded part is the
park. The diagonal lines show paths across the park. Each path is 1,120
feet long.

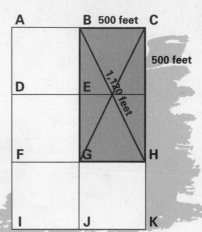

Example   Chan jogs along the same route each day. He
starts at point A and jogs to the park, point B.
He takes the path through the park to point H.
He returns home, passing through intersections
G, F, D, and back to A. How many total feet
does Chan jog each day?

▶ **Step 1.**   Trace Chan's route on the city map.

▶ **Step 2.**   Count the number of times he jogs a 500-foot
length. There are 5. Multiply by 500 feet to find
a total.

500 feet × 5 = 2,500

▶ **Step 3.**   Add the length of the diagonal path across the
park.

2,500 + 1,120 = 3,620

Chan jogs a total of 3,620 feet each day.

Solve.

1. Carolyn lives at intersection J. She jogs
   to the park, point G, across the park to
   point C, and back home again the same
   way. How many feet does Carolyn jog
   in one round trip?

2. Lynette is new to the city. She lives at
   intersection K. One day she walked to
   point B by going from point K to point
   H and across the park on the diagonal
   path. How many feet did she walk to
   get to point B?

Answer_____   Answer_____

3. Lynette walked home from point B by walking through intersections E, G, J, and then back home to K. How many feet did she walk in all, from point B back home?

4. Brad drives a delivery truck for the Pizza Parlor, which is at intersection G. One evening he needed to make stops at points E, C, A, and D. He went from the Pizza Parlor across the park to point C. Then he went to point A, and then to points D and E, and back to point G. How many total feet did Brad drive?

Answer_____

Answer_____

5. The distance around the outside edges of something is called the perimeter. Name the 4 corners of the perimeter of City Park.

6. How many feet is the perimeter of City Park?

Answer_____

Answer_____

7. Bob started walking at point I. He walked through intersections J, K, H, G, and F before returning to point I. Is the perimeter of City Park greater than the distance Bob walked?

8. What is the length of the shortest path going only once through every point on the map?
(Hint: Go from point I to A, from A to B, from B to J, from J to K, and from K to C.)

Answer_____

Answer_____

Two-Step Problems: Being a Consumer

The cost for summer concert tickets is shown in this sign.

OPERA TICKETS	
Adults	$ 10
Children (12 and under)	$ 5
Senior Citizens	$ 7
Adult Season Ticket (12 shows)	$100

Example Peter Chen is buying tickets for his family for one of the summer concerts. There are 2 adults and 3 children in his family. How much will it cost for Peter to take his family to a concert?

Step 1. To find the total cost for the adults and the total cost for the children, multiply.

$$\$10 \times 2 = \$20$$
$$\$5 \times 3 = \$15$$

Step 2. To find the total for the adults and the children, add.

$$\$20 + \$15 = \$35$$

It will cost Peter $35.

Solve.

1. The Chens want to buy season tickets. How much will the 2 adults save if they buy 2 season tickets instead of paying the adult prices for 12 concerts?

2. The Chen children have four grandparents, who are senior citizens. The grandparents attend 6 concerts each season. How much do the grandparents spend all together for the 6 concerts?

Answer_____   Answer_____

3. At its last performance, the concert hall sold a total of 2,527 tickets. Of these, 1,595 were adult tickets, 100 were season tickets, 530 were children's tickets, and the rest were senior citizens' tickets. How many senior citizens' tickets did they sell?

Answer_____

4. The concert hall has 3,000 seats. Of these, 100 seats are reserved for season ticket holders and 150 seats are reserved for families of the performers. How many seats are not reserved?

Answer_____

5. If the concert hall sold 1,926 adult tickets and 732 children's tickets, how much money did it make all together?

Answer_____

6. If the concert hall sold 342 tickets to senior citizens on Saturday and 418 on Sunday, how much money did it make on senior citizens' tickets?

Answer_____

7. The concert hall sold 1,263 children's tickets for an afternoon performance. How much more money would they have made if they had sold the same number of adult tickets?

Answer_____

8. On Friday night, the concert hall was sold out. They sold 1,900 adult tickets and 925 children's tickets. The rest were senior citizens' tickets. How many senior citizens' tickets did they sell?

Answer_____

Two-Step Problems: Finding An Average

Finding an average amount gives you a number that represents a group of numbers. Some examples are average rainfall, average miles per gallon, and average income.

To find an average, use two steps. First add the numbers in the group. Then divide the sum by the total number of items in the group.

Example The Berger's family income for 1987 to 1991 is listed in the chart. What was their average income for the 4 years?

Follow these steps.

Step 1.   Add the group of numbers.

Year	Income	Expenses
1987	$25,436	$10,200
1988	$26,506	$11,600
1989	$25,059	$ 9,072
1990	$27,275	$13,460

$$
\begin{array}{r}
\$25,436 \\
26,506 \\
25,059 \\
+\ 27,275 \\
\hline
\$104,276
\end{array}
$$

Step 2.   Divide the answer by the total number of items in the group.

$$
\begin{array}{r}
26,069 \\
4\overline{)104,276} \\
-\ 8 \\
\hline
24 \\
-\ 24 \\
\hline
0\ 27 \\
-\ 24 \\
\hline
36 \\
-\ 36 \\
\hline
0
\end{array}
$$

The Bergers' average income for 1987 to 1991 was $26,069.

Solve.

1. The chart shows the Bergers' expenses for 1987 to 1990. What was the yearly average for their expenses?

2. The Bergers' gas bills for the last 3 months were $80, $75, and $91. Find their average gas bill.

Answer_____   Answer_____

3. Mr. Berger is an auto mechanic. Last week he worked 10 hours on Monday, 8 hours on Tuesday, 9 hours on Wednesday, 11 hours on Thursday, and 7 hours on Friday. What was the average number of hours he worked each day?

Answer_____

4. Last month Mr. Berger worked 40 hours the first week, 42 hours the second week, 38 hours the third week, and 44 hours the fourth week. What was the average number of hours he worked each week?

Answer_____

5. Mrs. Berger works as a waitress. In tips last week, she made $28 on Monday, $30 on Tuesday, $29 on Wednesday, $33 on Thursday, and $45 on Friday. What was the average amount Mrs. Berger made in tips each day?

Answer_____

6. Last month Mrs. Berger earned $165, $155, $130, and $162 in tips. What was the average amount Mrs. Berger made in tips each week?

Answer_____

7. Mr. and Mrs. Berger are bowling in a 6-game tournament. Mr. Berger's scores were 300, 240, 240, 200, 270, and 280. What was his average score for the tournament?

Answer_____

8. Mrs. Berger bowled 300, 200, 250, 250, 280, and 250. What was her average score for the tournament?

Answer_____

Whole Numbers Skills Inventory

Write each number in words.

1. 32 _____

2. 246 _____

3. 2,316 _____

Compare each set of numbers. Write > or <.

4.
50 ☐ 40

5.
15 ☐ 25

6.
31 ☐ 13

7.
100 ☐ 69

Write the value of the underlined digit in each number.

8.
2̲6 _____

9.
3̲2 _____

10.
2̲51 _____

11.
5̲,480 _____

Round each number to the nearest ten.

12.
88 _____

13.
482 _____

14.
3,265 _____

Round each number to the nearest hundred.

15.
328 _____

16.
6,350 _____

17.
45,477 _____

Round each number to the nearest thousand.

18.
4,930 _____

19.
72,001 _____

20.
725,552 _____

Add.

21.
$5 + 9 =$

22.
$8 + 6 + 4 =$

23.
$$\begin{array}{r} 51 \\ + 20 \\ \hline \end{array}$$

24.
$6 + 13 + 50 =$

25.
$$\begin{array}{r} 210 \\ + 483 \\ \hline \end{array}$$

26.
$$\begin{array}{r} 3,240 \\ 425 \\ + \quad 33 \\ \hline \end{array}$$

27.
$$\begin{array}{r} 56 \\ + 98 \\ \hline \end{array}$$

28.
$85 + 22 + 60 =$

29.
$$\begin{array}{r} 492 \\ + 786 \\ \hline \end{array}$$

30.
$$\begin{array}{r} 432 \\ 3,297 \\ + 1,565 \\ \hline \end{array}$$

31.
$13,409 + 359,000 =$

32.
$$\begin{array}{r} 3,642,000 \\ 3,526 \\ + \quad 15,989 \\ \hline \end{array}$$

Subtract.

33.
$$\begin{array}{r} 15 \\ -\ 7 \\ \hline \end{array}$$

34. $12 - 4 =$

35.
$$\begin{array}{r} 35 \\ -24 \\ \hline \end{array}$$

36.
$$\begin{array}{r} 86 \\ -\ 3 \\ \hline \end{array}$$

37. $86 - 81 =$

38.
$$\begin{array}{r} 387 \\ -\ 26 \\ \hline \end{array}$$

39.
$$\begin{array}{r} 5,236 \\ -\ 102 \\ \hline \end{array}$$

40. $8,466 - 5,134 =$

41.
$$\begin{array}{r} 46 \\ -29 \\ \hline \end{array}$$

42. $73 - 9 =$

43.
$$\begin{array}{r} 30 \\ -\ 6 \\ \hline \end{array}$$

44.
$$\begin{array}{r} 60 \\ -35 \\ \hline \end{array}$$

45.
$$\begin{array}{r} 761 \\ -584 \\ \hline \end{array}$$

46. $835 - 276 =$

47. $4,200 - 2,198 =$

48.
$$\begin{array}{r} 15,060 \\ -\ 3,965 \\ \hline \end{array}$$

49.
$$\begin{array}{r} 240,000 \\ -\ 86,543 \\ \hline \end{array}$$

Multiply.

50. $6 \times 7 =$

51.
$$\begin{array}{r} 31 \\ \times\ 3 \\ \hline \end{array}$$

52. $501 \times 4 =$

53.
$$\begin{array}{r} 2,010 \\ \times\ \ \ 9 \\ \hline \end{array}$$

54.
$$\begin{array}{r} 31 \\ \times 12 \\ \hline \end{array}$$

55. $420 \times 43 =$

56.
$$\begin{array}{r} 210 \\ \times 231 \\ \hline \end{array}$$

57.
$$\begin{array}{r} 701 \\ \times 529 \\ \hline \end{array}$$

58.
$$\begin{array}{r} 7,100 \\ \times\ \ 100 \\ \hline \end{array}$$

59. $3,427 \times 1,000 =$

60. $32 \times 6 =$

61.
$$\begin{array}{r} 467 \\ \times\ \ 3 \\ \hline \end{array}$$

62.
$$\begin{array}{r} 4,208 \\ \times\ \ \ 7 \\ \hline \end{array}$$

63. $19 \times 27 =$

64. $9,040 \times 53 =$

65.
$$\begin{array}{r} 438 \\ \times 193 \\ \hline \end{array}$$

66.
$$\begin{array}{r} 306 \\ \times 700 \\ \hline \end{array}$$

67.
$$\begin{array}{r} 3,460 \\ \times\ \ 701 \\ \hline \end{array}$$

Divide.

68. $14 \div 7 =$ **69.** $5\overline{)40}$ **70.** $8\overline{)648}$ **71.** $6\overline{)366}$ **72.** $7\overline{)43}$ **73.** $23 \div 3 =$

74. $366 \div 5 =$ **75.** $8\overline{)1,744}$ **76.** $6\overline{)1,507}$ **77.** $2\overline{)10,480}$ **78.** $73\overline{)84}$

79. $64\overline{)610}$ **80.** $1,881 \div 19 =$ **81.** $50\overline{)40,420}$ **82.** $436\overline{)3,904}$ **83.** $841\overline{)85,796}$

Below is a list of the problems in this Skills Inventory and the pages on which the skills are taught. If you missed any problems, turn to the pages listed and practice the skills. Then correct the problems you missed in the Skills Inventory.

Problem	Practice Page	Problem	Practice Page	Problem	Practice Page
Unit 1		*Unit 3*		58-59	101-102
1-3	10	33-34	55-58	60-61	105-106
4-7	13	35-37	60-61	62	107-108
8-11	18-19	38-40	62	63-64	110-111
12-14	20	41-42	66-67	65	112-113
15-17	21	43-44	68-69	66-67	114-115
18-20	22	45-46	72-73	*Unit 5*	
Unit 2		47-49	75-78	68-69	123-126
21	29-31	*Unit 4*		70-71	127-128
22	32	50	85-87	72-74	129-130
23-24	34-35	51	89-90	75	134-135
25	36	52-53	91	76-77	136-137
26	37	54	95	78-80	140-142
27-28	41-42	55	96	81	143
29	43	56-57	97-98	82	147-148
30	44			83	149-150
31	45				
32	48				

Glossary

addend (page 31) - A number that you add in an addition problem.

$$\begin{array}{r} 5 \\ + 3 \\ \hline 8 \end{array}$$

addition (page 27) - Putting numbers together to find a total. The symbol + is used in addition.

$$\begin{array}{r} 6 \\ + 7 \\ \hline 13 \end{array}$$

average (page 164) - The amount you get when you divide a total by the number of items you added to get that total.

$$\left.\begin{array}{r} 3 \\ 4 \\ + 8 \\ \hline 15 \end{array}\right\} 3 \text{ items} \qquad 3\overline{)15}^{\,5}$$

bar graph (page 159) - A graph with bars of different lengths that stand for certain numbers.

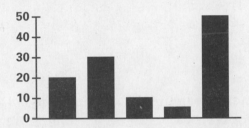

borrowing (page 66) - Taking an amount from a top digit in subtraction and adding it to the next digit to the right.

$$\begin{array}{r} {}^{1}\;{}^{16} \\ \cancel{2}\;\cancel{6} \\ - \quad 9 \\ \hline 1\;7 \end{array}$$

carrying (page 44) - Taking an amount from the sum of digits with the same place value and adding it to the next column of digits to the left.

$$\begin{array}{r} 1 \\ 18 \\ + 7 \\ \hline 25 \end{array}$$

circle graph (page 93) - A circle cut into sections to show the parts that make a total.

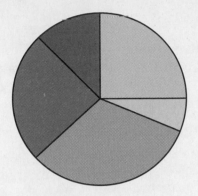

column (page 30) - A vertical line of numbers.

$$\begin{array}{ccc} 3 & 2 & 9 \\ 5 & 4 & 1 \\ 0 & 6 & 7 \end{array}$$

comparing (page 13) - Deciding if a number is equal to, greater than, or less than another number.

difference (page 57) - The answer to a subtraction problem.

$$\begin{array}{r} 15 \\ - 10 \\ \hline 5 \end{array}$$

digit (page 12) - One of the ten symbols used to write numbers.

0 1 2 3 4 5 6 7 8 9

dividend (page 125) - The number that you divide into in a division problem.

$$6\overline{)30}^{\,5}$$

division (page 121) - Splitting an amount into equal groups. The symbols ÷ and $\overline{)}$ are used in division.

$$12 \div 2 = 6 \qquad 2\overline{)12}^{\,6}$$

divisor (page 125) - The number you divide by in a division problem.

$$6\overline{)30}^{\,5}$$

equal (page 15) - The same in value. The symbol = means *equal*.

6 × 6 = 36

estimating (page 38) - Finding an answer by rounding the numbers in a problem. You use estimating when an exact answer is not needed.

greater than (page 13) - More than. The symbol > means *greater than*.

6 > 5 means 6 is greater than 5

horizontal (page 31) - Side-to-side.

---

hundred (page 10) - The word name for 100.

lead digit (page 49) - The first digit on the left in a number with two or more digits.

1,365

less than (page 13) - Smaller than. The symbol < means *less than*.

7 < 12 means 7 is less than 12

million (page 47) - The word name for 1,000,000.

minuend (page 57) - The number that you subtract from in a subtraction problem.

$$\begin{array}{r} 15 \\ -\ 10 \\ \hline 5 \end{array}$$

multiplicand (87) - The top number in a multiplication problem.

$$\begin{array}{r} 54 \\ \times\ 3 \\ \hline 162 \end{array}$$

multiplication (page 83) - Combining equal numbers two or more times to get a total. The symbol × is used in multiplication.

$$\begin{array}{r} 54 \\ \times\ 3 \\ \hline 162 \end{array}$$

multiplier (page 87) - The bottom number in a multiplication problem.

$$\begin{array}{r} 54 \\ \times\ 3 \\ \hline 162 \end{array}$$

number line (page 9) - A line with equally spaced points that are labeled with numbers.

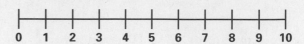

one (page 10) - The word name for 1.

operation (page 155) - The process you use to solve a math problem. The basic operations are addition, subtraction, multiplication, and division.

partial product (page 95) - The total you get when you multiply a number by one digit of another number.

$$\begin{array}{r} 13 \\ \times\ 22 \\ \hline 26 \\ +\ 26 \\ \hline 286 \end{array}$$

perimeter (page 161) - The distance around the outside edges of a figure.

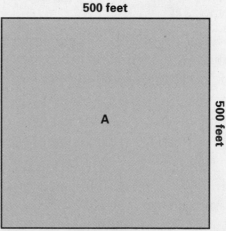

Perimeter of block A is
500 feet x 4 sides = 2000 feet

pictograph (page 104) - A graph that uses symbols or pictures to stand for certain numbers.

A	B	C	D
⛳	⛳		⛳
🎿		🎿	
⋂		⋂	⋂
🚲		🚲	🚲
	🚶🚶	🚶🚶	
	🎋		🎋
	🐟		🐟

place value (page 18) - The value of a digit, based on its place in a number.

The value of 9 in the number 590 is **90**

plus (page 41) - To add. The symbol for plus is +.

product (page 87) - The answer to a multiplication problem.

$$\begin{array}{r} 54 \\ \times\ 3 \\ \hline 162 \end{array}$$

quotient (page 125) - The answer to a division problem.

$$6\overline{)30}\quad 5$$

rectangle (page 64) - A four-sided figure with equal opposite sides.

remainder (page 129) - The amount left over in a division problem.

$$\begin{array}{r} 6\ \textbf{R1} \\ 6\overline{)37} \\ -36 \\ \hline 1 \end{array}$$

renaming (page 41) - Carrying or borrowing a number.

$$\begin{array}{r} 1 \\ 26 \\ +\ 9 \\ \hline 35 \end{array} \qquad \begin{array}{r} 1\ 16 \\ 2\,6 \\ -\ 9 \\ \hline 17 \end{array}$$

rounding (page 20) - Expressing a number to the nearest ten, hundred, thousand, and so on.

row (page 29) - A horizontal line of numbers.

3 2 9
5 4 1
0 6 7

subtraction (page 53) - Taking away a certain amount from another amount to find a difference. The symbol − is used in subtraction.

$$\begin{array}{r} 15 \\ -10 \\ \hline 5 \end{array}$$

subtrahend (page 57) - The number that you subtract in a subtraction problem.

$$\begin{array}{r} 15 \\ -\underline{10} \\ 5 \end{array}$$

sum (page 31) - The answer to an addition problem.

$$\begin{array}{r} 5 \\ +\ 3 \\ \hline 8 \end{array}$$

table (page 23) - Information arranged in rows and columns.

	1980	1990
Pattonville	12,965	13,012
Shoreline	23,312	23,501
Eagle City	11,573	11,416
Benton	24,599	24,467
Hillview	22,207	22,299

ten (page 10) - The word name for 10.

thousand (page 10) - The word name for 1,000.

times (page 85) - To multiply. The symbol for times is ×.

vertical (page 31) - Up-and-down.

zero (page 10) - The word name for 0.

Answers & Explanations

The answer to the problem that was worked out for you in the lesson is written here in blue. The next answer has an explanation written beneath it. The answers to the rest of the problems in the lesson follow in order.

Skills Inventory

Page 6
1. twenty-five
2. one hundred seventy-seven
3. three thousand, five hundred eleven
4. 60 < 70
5. 25 > 15
6. 41 > 14
7. 89 < 100
8. 10
9. 7
10. 100
11. 2,000
12. 60
13. 670
14. 1,360
15. 400
16. 7,900
17. 33,800
18. 9,000
19. 45,000
20. 235,000
21. 11
22. 20
23. 77
24. 99
25. 675
26. 5,999
27. 91
28. 169
29. 914
30. 7,861
31. 280,677
32. 2,458,700

Page 7
33. 7
34. 8
35. 24
36. 71
37. 6
38. 662
39. 8,120
40. 3,126
41. 9
42. 73
43. 33
44. 15
45. 148
46. 467
47. 3,814
48. 10,078
49. 30,531
50. 24
51. 104
52. 2,706
53. 12,080
54. 299
55. 16,960
56. 13,530
57. 219,876
58. 610,000
59. 2,409,000
60. 175
61. 1,556
62. 17,545
63. 570
64. 279,220
65. 48,411
66. 305,400
67. 383,540

Page 8
68. 2
69. 6
70. 71
71. 91
72. 7 R3
73. 5 R1
74. 96 R3
75. 328 R2
76. 217 R1
77. 5,115
78. 1 R7
79. 9 R18
80. 77
81. 688 R10
82. 9 R528
83. 103 R1

Unit 1

Page 9
1. 7
2. 15

   If you count each mark until you reach "B," you will have counted to 15.
3. 28
4. 33
5. 49

Page 10

zero	one
two	three
four	five
six	seven
eight	nine
ten	eleven
twelve	thirteen
fourteen	fifteen
sixteen	seventeen
eighteen	nineteen
twenty	twenty-one
twenty-two	twenty-three
twenty-four	twenty-five
thirty	forty
fifty	sixty
seventy	eighty
ninety	one hundred
one thousand	

Page 11
1. 14

   count 3 more
2. 20

   count 5 more Each number is 5 more than the last number. The next number would be 20.
3. 20

   count 4 more
4. 7

   count 2 less
5. 0

   count 3 less
   Each number is 3 less than the last number. The next number would be 0.

6. 8
    count 3 less

7. 30
    count 4 more

8. 21
    count 2 less

9. 12
    count 7 less

10. 22
    count 6 more

11. 9
    count 5 less

12. 1
    count 3 less

Page 12

1. 2 ones

2. 6 ones
    The number 6 has six ones.

3. 1 tens 5 ones

4. 8 tens 9 ones
    The number 89 has 8 tens and 9 ones.

5. 1 tens 0 ones

6. 2 tens 0 ones

7. 5 tens 0 ones

8. 9 tens 0 ones

9. 2 hundreds
    5 tens 4 ones

10. 9 hundreds 7 tens 1 ones
    The number 971 has 9 hundreds, 7 tens, and 1 ones.

11. 8 hundreds
    5 tens 5 ones

12. 4 hundreds
    8 tens 2 ones

13. 1 hundreds
    0 tens 6 ones

14. 6 hundreds
    0 tens 5 ones

15. 4 hundreds
    0 tens 8 ones

16. 3 hundreds
    0 tens 2 ones

17. 5 hundreds
    0 tens 0 ones

18. 1 hundreds
    0 tens 0 ones

19. 7 hundreds
    0 tens 0 ones

20. 6 hundreds
    0 tens 0 ones

Page 13

1. 50 > 40
2. 10 < 20
3. 80 > 10
    80 is to the right of 10 on the number line.
4. 70 > 30
5. 25 < 30
6. 51 > 40
7. 10 < 29
8. 30 < 42
9. 100 > 96
10. 52 > 34
11. 48 < 54
12. 64 < 68
13. 79 < 82
14. 63 < 83
15. 91 > 81
16. 22 < 33
17. 41 > 22
18. 33 < 35
19. 100 > 10
20. 89 < 99
21. 42 > 24
22. 89 < 98
23. 19 < 91
24. 0 < 10

Page 14

Answers may vary.

1. **2 and 3 more is 5**
    6 and 1 less is 5

2. 4 and 4 more is 8
    10 and 2 less is 8

3. 8 and 2 more is 10
    15 and 5 less is 10

4. 12 and 2 more is 14
    15 and 1 less is 14

Page 15

1. a
2. b
3. a
4. a
5. b
6. a
7. b
8. a
9. a
10. a
11. b
12. b

Page 16

1. 2
    count 2 less

2. 9
    count 3 less

3. 26
    count 5 more

4. 28
    count 7 more

5. 6
    count 6 less

6. 30
    count 7 less

7. 4 ones
8. 2 tens 8 ones

9. 1 hundreds
    3 tens 4 ones

10. 6 hundreds
    1 tens 4 ones

11. 25 < 31
12. 42 > 22
13. 37 < 41
14. 20 < 50
15. 44 > 39
16. 51 > 15

17.–20. Answers may vary.

17. 3 and 1 more is 4
    7 and 3 less is 4

18. 5 and 2 more is 7
    10 and 3 less is 7

19. 6 and 5 more is 11
    15 and 4 less is 11

20. 9 and 4 more is 13
    15 and 2 less is l3

21. b
22. a

Page 17

1. last week
    36 > 32

2. this week
    Compare 34 and 43. 43 > 34

3. Rosa
4. Rosa
5. last Sunday
6. this Sunday
7. greater
8. last year

Page 18

1. 6 tens = 60
    7 ones = 7

2. 9 tens = 90
    2 ones = 2
    The number 92 has 9 tens, or 90; and 2 ones, or 2.

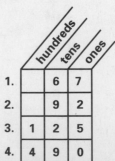

	hundreds	tens	ones
1.		6	7
2.		9	2
3.	1	2	5
4.	4	9	0

3. 1 hundreds = 100
2 tens = 20
5 ones = 5

4. 4 hundreds = 400
9 tens = 90
0 ones = 0
The number 490 has 4 hundreds, or 400;
9 tens, or 90; and 0 ones, or 0.

5. 50   **6.** 6
The value of 6 in the ones place is 6.

7. 90   **8.** 100
9. 70   **10.** 4
11. 200   **12.** 300
13. 70   **14.** 9
15. 80   **16.** 0

Page 19

1. 2 thousands = 2,000
4 hundreds = 400
3 tens = 30
7 ones = 7

2. 3 thousands = 3,000
0 hundreds = 0
1 tens = 10
3 ones = 3
The number 3,013 has 3 thousands, or
3,000; 0 hundreds, or 0; 1 tens, or 10; and 3
ones, or 3.

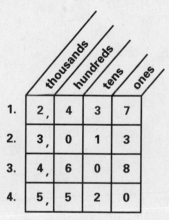

3. 4 thousands = 4,000
6 hundreds = 600
0 tens = 0
8 ones = 8

4. 5 thousands = 5,000
5 hundreds = 500
2 tens = 20
0 ones = 0

5. 500
6. 90
The value of 9 in the tens place is 90.

7. 3   **8.** 30
9. 3,000   **10.** 8,000
11. 400   **12.** 2
13. 50

Page 20

1. 50
2. 60
8 > 5, so 58 rounds up to 60.
3. 50   **4.** 50
5. 70   **6.** 60
7. 890
8. 910
1 < 5, so 911 rounds down to 910.
9. 450   **10.** 790
11. 390   **12.** 180
13. 9,550
14. 1,410
6 > 5, so 1,406 rounds up to 1,410.
15. 6,500   **16.** 2,000
17. 8,770   **18.** 1,980
19. 40   **20.** 220
21. 900   **22.** 1,890
23. 80   **24.** 9,990

Page 21

1. 400
2. 500
4 < 5, so 540 rounds down to 500.
3. 700   **4.** 600
5. 700   **6.** 400
7. 1,600
8. 2,600
4 < 5, so 2,640 rounds down to 2,600.
9. 4,700   **10.** 1,200
11. 1,100   **12.** 2,100
13. 16,700
14. 24,400
4 < 5, so 24,441 rounds down to 24,400.
15. 92,600   **16.** 18,500
17. 10,400   **18.** 13,900
19. 800   **20.** 50,100
21. 9,800   **22.** 2,600
23. 33,800   **24.** 600

Page 22

1. 7,000
2. 6,000
5 = 5, so 5,500 rounds up to 6,000.
3. 4,000   **4.** 4,000
5. 6,000   **6.** 5,000
7. 18,000
8. 98,000
9 > 5, so 97,999 rounds up to 98,000.

9. 37,000   **10.** 22,000
11. 14,000   **12.** 92,000
13. **112,000**
14. 214,000
    6 > 5, so 213,649 rounds up to 214,000.
15. 139,000   **16.** 265,000
17. 956,000   **18.** 231,000
19. 9,000   **20.** 149,000
21. 47,000   **22.** 561,000
23. 86,000   **24.** 7,000

Page 23

1. Eagle City   **2.** Pattonville, Shoreline, and Hillview

3. Eagle City and Benton   **4.** Shoreline

Page 24

5. Eagle City   **6.** Benton
7. 1990   **8.** Eagle City
9. Hillview   **10.** 1990
11.

City	1980 Population	Nearest Thousand	1990 Population	Nearest Thousand
Pattonville	12,965	13,000	13,012	13,000
Shoreline	23,312	23,000	23,501	24,000
Eagle City	11,573	12,000	11,416	11,000
Benton	24,599	25,000	24,467	24,000
Hillview	22,207	22,000	22,299	22,000

Unit 1 Review, page 25

1. twenty-five
2. forty-one
3. nine hundred eighty-seven
4. six hundred three
5. one thousand one
6. six thousand, eight hundred fifty-two
7. 65   **8.** 24
    count 10 more   count 2 more
9. 12   **10.** 9
    count 3 more   count 2 less
11. 50   **12.** 8
    count 10 more   count 2 less
13. 20 > 19   **14.** 44 < 72
15. 11 > 10   **16.** 38 < 98
17. 90 > 70   **18.** 64 > 40
19. 44 > 33   **20.** 22 < 26

21.–24. Answers may vary.

21. 4 and 2 more is 6   **22.** 1 and 1 more is 2
    7 and 1 less is 6   4 and 2 less is 2

23. 6 and 6 more is 12   **24.** 5 and 4 more is 9
    15 and 3 less is 12   12 and 3 less is 9

Page 26

25. a   **26.** b
27. b   **28.** a
29. 100   **30.** 20
31. 900   **32.** 8,000
33. 7   **34.** 40
35. 80   **36.** 20
37. 60   **38.** 100
39. 160   **40.** 270
41. 1,240   **42.** 4,560
43. 500   **44.** 900
45. 600   **46.** 200
47. 1,200   **48.** 5,900
49. 33,500   **50.** 58,500
51. 7,000   **52.** 4,000
53. 9,000   **54.** 11,000
55. 35,000   **56.** 99,000
57. 168,000   **58.** 133,000

Unit *2*

Page 27

1. 3 tens 7 ones
2. 5 hundreds
    5 tens 0 ones
    The number 550 has 5 hundreds, 5 tens, and 0 ones.
3. 3 thousands
    9 hundreds
    0 tens 1 ones

Page 28

4. 2,000
5. 5
    The value of 5 in the ones place is 5.
6. 40   **7.** 800
8. 200   **9.** 50
10. 9   **11.** 0
12. 9,000   **13.** 8
14. 40
15. 120
    2 < 5, so 122 rounds down to 120.
16. 310   **17.** 2,450
18. 100
19. 4,600
    5 = 5, so 4,559 rounds up to 4,600.
20. 16,600   **21.** 100
22. **1,000**
23. 16,000
    9 > 5, so 15,987 rounds up to 16,000.
24. 234,000   **25.** 1,000

Page 29

1.	0	1	2	3	4	5	6	7	8	9
2.	1	2	3	4	5	6	7	8	9	10
3.	2	3	4	5	6	7	8	9	10	11
4.	3	4	5	6	7	8	9	10	11	12
5.	4	5	6	7	8	9	10	11	12	13
6.	5	6	7	8	9	10	11	12	13	14
7.	6	7	8	9	10	11	12	13	14	15
8.	7	8	9	10	11	12	13	14	15	16
9.	8	9	10	11	12	13	14	15	16	17
10.	9	10	11	12	13	14	15	16	17	18

Page 30

+	0	1	2	3	4	5	6	7	8	9
0	0	1	2	3	4	5	6	7	8	9
1	1	2	3	4	5	6	7	8	9	10
2	2	3	4	5	6	7	8	9	10	11
3	3	4	5	6	7	8	9	10	11	12
4	4	5	6	7	8	9	10	11	12	13
5	5	6	7	8	9	10	11	12	13	14
6	6	7	8	9	10	11	12	13	14	15
7	7	8	9	10	11	12	13	14	15	16
8	8	9	10	11	12	13	14	15	16	17
9	9	10	11	12	13	14	15	16	17	18

1. 9

2. 12

Find 3 in the column along the top. Find the row beginning with 9. Move right on the row until it meets the column with 3. The place where the row and column meet is the sum of 3 + 9, or 12.

3.	12	**4.**	2
5.	13	**6.**	11
7.	12	**8.**	14
9.	2	**10.**	10
11.	13	**12.**	11
13.	9	**14.**	14
15.	9	**16.**	5
17.	16	**18.**	10
19.	18	**20.**	10

Page 31

1.	14	**2.**	10
3.	16	**4.**	9
5.	1	**6.**	2
7.	6	**8.**	6
9.	0	**10.**	3
11.	8	**12.**	6
13.	12	**14.**	9
15.	0	**16.**	11
17.	6	**18.**	16

19.	1	**20.**	9
21.	10	**22.**	5
23.	4	**24.**	5

25.–40. Answers may vary.

25.
5 + 4 = 9
8 + 1 = 9
7 + 2 = 9
6 + 3 = 9

26.
6 + 9 = 15
7 + 8 = 15
8 + 7 = 15
9 + 6 = 15

27.
9 + 9 = 18
12 + 6 = 18
10 + 8 = 18
11 + 7 = 18

28.
5 + 7 = 12
6 + 6 = 12
7 + 5 = 12
8 + 4 = 12

29.
4 + 1 = 5
3 + 2 = 5
2 + 3 = 5
1 + 4 = 5

30.
6 + 7 = 13
7 + 6 = 13
8 + 5 = 13
9 + 4 = 13

31.
5 + 2 = 7
4 + 3 = 7
3 + 4 = 7
2 + 5 = 7

32.
4 + 0 = 4
3 + 1 = 4
2 + 2 = 4
1 + 3 = 4

33.
7 + 9 = 16
9 + 7 = 16
8 + 8 = 16
10 + 6 = 16

34.
5 + 5 = 10
3 + 7 = 10
4 + 6 = 10
8 + 2 = 10

35.
6 + 8 = 14
7 + 7 = 14
8 + 6 = 14
9 + 5 = 14

36.
9 + 2 = 11
8 + 3 = 11
7 + 4 = 11
6 + 5 = 11

37.
9 + 8 = 17
8 + 9 = 17
10 + 7 = 17
7 + 10 = 17

38.
7 + 1 = 8
6 + 2 = 8
5 + 3 = 8
4 + 4 = 8

39.
5 + 1 = 6
4 + 2 = 6
3 + 3 = 6
2 + 4 = 6

40.
2 + 1 = 3
0 + 3 = 3
1 + 2 = 3
3 + 0 = 3

Page 32

1. 11

2. 10

Add the first two digits. 6 + 2 = 8.
Then add the last digit to the 8. 2 + 8 = 10.

$$\begin{array}{r} 6 \\ 2 \\ +\ 2 \\ \hline 10 \end{array}$$

3.	9	**4.**	10
5.	13	**6.**	11

7. 11

8. 6

Line up the digits in a column. Add the first two digits. 3 + 2 = 5. Then add the 1 to 5. 1 + 5 = 6.

$$\begin{array}{r} 3 \\ 2 \\ +\ 1 \\ \hline 6 \end{array}$$

9. 12   **10.** 13
11. 18
12. 21
Line up the digits. Add the first two digits. 7 + 3 = 10. Add the last 2 digits. 9 + 2 = 11. Add the two sums. 10 + 11 = 21.

$$\begin{array}{r} 7 \\ 3 \\ 9 \\ +\ 2 \\ \hline 21 \end{array}$$

13. 17   **14.** 21
15. 15 hours
Add to find the total hours Roy worked.

$$\begin{array}{r} 4 \\ 6 \\ +\ 5 \\ \hline 15 \end{array}\ \text{hours}$$

16. 18 tomatoes
Add to find how many tomatoes Jean grew in all.

$$\begin{array}{r} 3 \\ 6 \\ 4 \\ +\ 5 \\ \hline 18 \end{array}\ \text{tomatoes}$$

Page 33

1. c
The beach umbrella costs $8. The T-shirt costs $4. These two amounts added together, 8 + 4 = 12, will be less than the $15 Harry has to spend. He will get change from $15.

2. b
From the sign, you can see that a beach towel costs $5. If she buys a towel and a skirt, the cost will be greater than $10. If she buys a towel and cutoffs, the cost will also be greater than $10.

3. a
The total bill for the three items Jasmine plans to buy would be $15. 7 + 5 + 3 = 15. 15 > 10.

4. a
The two items Carl bought cost $9 each. 9 + 9 = 18. The four items Kai bought cost $16 all together. 18 > 16.

5. c
The cost of the two items Chris wants to buy is $16. 7 + 9 = 16. The only item he would have enough money to buy is the sandals. 16 + 3 = 19. 19 < 20.

6. b
The two items Sue bought cost $11. 4 + 7 = 11. The only other two items that equal $11 would be the skirt, $6, and a beach towel, $5. 6 + 5 = 11.

Page 34

1. 84
2. 77
Add the ones. 5 + 2 = 7. Then add the tens. 6 + 1 = 7.

$$\begin{array}{r} 65 \\ +\ 12 \\ \hline 77 \end{array}$$

3. 99   **4.** 67
5. 92   **6.** 48
7. 96
8. 48
Add the ones. 5 + 3 = 8. Then add the tens. 4 + 0 = 4.

$$\begin{array}{r} 45 \\ +\ 3 \\ \hline 48 \end{array}$$

9. 18   **10.** 58
11. 39   **12.** 18
13. 99   **14.** 91
15. 89   **16.** 64
17. 38   **18.** 67
19. 87   **20.** 29
21. 35   **22.** 88
23. 78   **24.** 45
25. 81   **26.** 69
27. 79   **28.** 58
29. 95   **30.** 77

Page 35

1. 69
2. 28
Line up the digits. Then add the ones. 5 + 3 = 8. Add the tens. 1 + 1 = 2.

$$\begin{array}{r} 15 \\ +\ 13 \\ \hline 28 \end{array}$$

3. 95   **4.** 84
5. 68   **6.** 49
7. 77   **8.** 44
9. 49   **10.** 78

11. 67 **12.** 48
Line up the digits. Then add the ones.
2 + 4 + 1 = 7. Add the tens. 3 + 2 + 1 = 6.

```
  32
  24
+ 11
  67
```

13. 69 **14.** 19 miles in all

15. $13

```
   10
    5
 + 4
   19
```

```
$13
  1
+ 12
 13
```

Page 36

1. 129

2. 899
Add the ones. 4 + 5 = 9. Add the tens.
9 + 0 = 9. Then add the hundreds 7 + 1 = 8.

```
  794
+ 105
  899
```

3. 174 **4.** 997
5. 435 **6.** 799
7. 538 **8.** 379
9. 398 **10.** 218
11. 589
12. 899
First add the ones. 2 + 0 + 7 = 9. Add the
tens. 2 + 0 + 7 = 9. Add the hundreds.
6 + 1 + 1 = 8.

```
  622
  100
+ 177
  899
```

13. 368 **14.** 597
15. 367 **16.** 469
17. 895 **18.** 888
Add the ones. 1 + 3 + 0 + 1 = 5. Add the
tens. 0 + 8 + 1 + 0 = 9. Add the hundreds.
2 + 3 + 3 = 8.

```
  201
  383
   10
+ 301
  895
```

19. 738 **20.** 684

Page 37

1. 1,589
2. 3,800
Line up the digits. Add the ones.
0 + 0 = 0. Add the tens. 0 + 0 = 0. Add the
hundreds. 3 + 5 = 8. Add the thousands.
3 + 0 = 3.

```
  3,300
+   500
  3,800
```

3. 5,995 **4.** 6,936
5. 1,755 **6.** 2,589
7. 7,054 **8.** 5,367
9. 2,569 **10.** 5,979
11. 9,967 **12.** 7,885

Page 38

1. $80 **2.** $50
$13 rounds to $10 $10 + $10 + $10
$22 rounds to $20 + $20 = $50
$49 rounds to $50
$10 + $20 + $50 = $80

Page 39

3. $80
$10 + $20 + $40 + $10 = $80
4. Clothes
$80 > $50
5. Food
$230 > $80
6. $780
$500 + $230 + $50 = $780
7. $1,080
$1,000 + $80 = $1,080
8. $70
$50 + $20 = $70
9. $260 **10.** No
$240 + $20 = $260 $250 < $260

Page 40

1. 5 **2.** 70
3. 9 **4.** 50
5. 700 **6.** 900
7. 1,000 **8.** 700
9. 9,000 **10.** 60
11. 40 **12.** 80
13. 150 **14.** 980
15. 820 **16.** 2,470
17. 5,310 **18.** 4,990
19. 68 **20.** 34
21. 78 **22.** 89
23. 99 **24.** 99
25. 66 **26.** 74
27. 79 **28.** 97
29. 4,875 **30.** 4,867
31. 9,978 **32.** 9,889
33. 4,798 **34.** 9,963
35. 9,579

Page 41

1. 42

2. 54

Line up the digits. Add the ones. 7 + 7 = 14. Rename 14 ones as 1 ten and 4 ones. Write 4 in the ones column. Carry 1 ten. Add the tens. 1 + 2 + 2 = 5.

```
  1
  27
+ 27
  54
```

3. 90   **4.** 90
5. 80   **6.** 65
7. 80   **8.** 92
9. 30   **10.** 70
11. 97   **12.** 75
13. 92   **14.** 95
15. 85   **16.** 44
17. 83   **18.** 33
19. 43   **20.** 76

Page 42

1. 115
2. 103

Add the ones. 4 + 9 = 13. Rename. Write 3 in the ones column. Carry 1 ten. Add the tens. 1 + 7 + 2 = 10. Write 0 in the tens column. Write 1 in the hundreds column.

```
  1
  74
+ 29
 103
```

3. 120   **4.** 171
5. 144   **6.** 143
7. 129   **8.** 142
9. 152   **10.** 166
11. 174   **12.** 124
13. 200   **14.** 198
15. 157   **16.** 123
17. 139   **18.** 96

Page 43

1. 901
2. 903

Add the ones. 8 + 5 = 13. Carry 1 ten. Add the tens. 1 + 7 + 2 = 10. Carry 1 hundred. Add the hundreds. 1 + 8 = 9.

```
  11
  878
+  25
  903
```

3. 1,012   **4.** 1,020
5. 1,000   **6.** 916
7. 915   **8.** 1,106
9. 451   **10.** 1,192

11. 1,592

Add the ones. 1 + 8 + 3 = 12. Carry 1 ten. Add the tens. 1 + 3 + 2 + 3 = 9. Add the hundreds. 3 + 8 + 4 = 15.

```
   1
  331
  828
+ 433
1,592
```

12. 2,641   **13.** 888
14. 2,434   **15.** 2,409
16. 1,014   **17.** 1,005
18. 1,633

Page 44

1. 2,227
2. 2,743

Add the ones. 9 + 4 = 13. Carry 1 ten. Add the tens. 1 + 7 + 6 = 14. Carry 1 hundred. Add the hundreds. 1 + 5 + 1 = 7. Add the thousands. 2 + 0 = 2

```
   11
 2,579
+  164
 2,743
```

3. 1,202   **4.** 1,220
5. 3,542   **6.** 5,477
7. 2,224   **8.** 9,909
9. 16,121   **10.** 19,434
11. 18,199   **12.** 17,987
13. 30,510   **14.** 51,221
15. 18,558   **16.** 71,519

Page 45

1. 410
2. 710

Add the ones. 9 + 1 = 10. Carry 1 ten. Add the tens. 1 + 0 + 0 = 1. Add the hundreds. 4 + 3 = 7.

```
  1
  409
+ 301
  710
```

3. 1,311   **4.** 1,912
5. 6,710   **6.** 5,164
7. 6,213   **8.** 992
9. 7,012   **10.** 9,817
11. 18,118   **12.** 26,917
13. 10,210   **14.** 14,610
15. 23,414   **16.** 48,514

Page 46

1. 29

2. 27

Add groups of digits in the ones column.
$7 + 8 = 15.$ $5 + 5 + 2 = 12.$ $15 + 12 = 27.$

```
   7
   8
   5
   5
 + 2
  27
```

3.	240	**4.**	223
5.	1,746	**6.**	1,092
7.	8,108	**8.**	15,788
9.	22,023	**10.**	14,722
11.	5,385	**12.**	69,320

Page 47

1.

3 millions	3,000,000
4 hundred thousands	400,000
7 ten thousands	70,000
2 thousands	2,000
4 hundreds	400
0 tens	00
0 ones	0

2.

6 millions	6,000,000
4 hundred thousands	400,000
8 ten thousands	80,000
0 thousands	0,000
6 hundreds	600
0 tens	00
3 ones	3

3. 40

4. 3

The value of 3 ones in 103 is 3.

5.	4,000	**6.**	900
7.	20,000	**8.**	6,000
9.	200,000	**10.**	3,000
11.	3,000,000		

Page 48

1. **119,541**

2. 145,278

Add, beginning with the ones place.

```
     11
   38,468
 + 106,810
  145,278
```

3.	157,142	**4.**	1,823,032
5.	1,499,000	**6.**	796,846
7.	63,340	**8.**	2,160,614
9.	3,049,382	**10.**	2,267,541
11.	7,834,034	**12.**	1,372,497
13.	61,421	**14.**	149,403
15.	347,710		

Page 50

1. 600 miles   **2.** 800 miles

180 miles rounds to	100
200 miles	100
193 miles rounds to	200
200 miles	200
245 miles rounds to	+ 200
200 miles	800 miles

$200 + 200 + 200 = 600$ miles

3. 80 gallons
$10 + 10 + 20 + 20 + 20 = 80$ gallons

4. $390
$60 + $80 + $90 + $100 + $60 = $390

5. less
$390 < $500

6. 2,000 miles
1,927 miles rounds to 2,000 miles

7. last year
2,000 > 800

8. 3,000 miles
2,794 miles rounds to 3,000 miles

9. 6,000 miles
$3,000 + 3,000 = 6,000$ miles

10. next year
800 < 6,000

Unit 2 Review, page 51

1.	11	**2.**	10
3.	11	**4.**	12
5.	18	**6.**	10
7.	99	**8.**	95
9.	95	**10.**	99
11.	89	**12.**	99
13.	64	**14.**	97
15.	68	**16.**	89
17.	349	**18.**	899
19.	449	**20.**	1,997
21.	6,898	**22.**	6,297
23.	5,899	**24.**	2,555
25.	90	**26.**	95
27.	140	**28.**	95
29.	105	**30.**	120
31.	107	**32.**	54
33.	145	**34.**	163

Page 52

35.	315	**36.**	10,003
37.	10,120	**38.**	1,100
39.	1,384	**40.**	943
41.	4,029	**42.**	2,163
43.	9,817	**44.**	18,118
45.	26,917	**46.**	1,746
47.	1,092	**48.**	106,056
49.	1,088,033	**50.**	7,745,811

51. 12,173,634   **52.** 2,435,355
53. 58,767   **54.** 601,175

Unit 3

Page 53

1. 9 ten thousands   **2.** 4 ten thousands
    3 thousands   2 thousands
    4 hundreds   0 hundreds
    2 tens   7 tens
    7 ones   6 ones

Page 54

3. 498   **4.** 375

$$\begin{array}{r} 1 \\ 346 \\ +\ 29 \\ \hline 375 \end{array}$$

5. 5,281   **6.** 9,934
7. 2,533   **8.** 8,731
9. 1,529   **10.** 2,024
11. 30   **12.** 160

                5 = 5, so 155
                rounds up to 160

13. 410   **14.** 1,300
15. 400   **16.** 800
17. 1,000   **18.** 5,400
19. 6,000   **20.** 9,000
21. 15,000   **22.** 100,000
23. 300   **24.** 28

                28 >18

25. 73   **26.** 297
27. 860   **28.** 4,000

Page 55

1.	0	1	2	3	4	5	6	7	8	9
2.	0	1	2	3	4	5	6	7	8	9
3.	0	1	2	3	4	5	6	7	8	9
4.	0	1	2	3	4	5	6	7	8	9
5.	0	1	2	3	4	5	6	7	8	9
6.	0	1	2	3	4	5	6	7	8	9
7.	0	1	2	3	4	5	6	7	8	9
8.	0	1	2	3	4	5	6	7	8	9
9.	0	1	2	3	4	5	6	7	8	9
10.	0	1	2	3	4	5	6	7	8	9

Page 56

1. 6
2. 7

Find 8 in the farthest column on the left.
Move right on the row to 15. Move up from
15 to the top of the column to find 7.

3. 8   **4.** 4
5. 9   **6.** 5
7. 5   **8.** 3

9. 6   **10.** 3
11. 9   **12.** 9
13. 2   **14.** 8
15. 5   **16.** 8
17. 16   **18.** 0

Page 57

1. 8   **2.** 4
3. 4   **4.** 6
5. 4   **6.** 8
7. 7   **8.** 0
9. 11   **10.** 11
11. 10   **12.** 10
13. 4   **14.** 5
15. 5   **16.** 6
17. 15   **18.** 9
19. 7   **20.** 7
21. 5   **22.** 2
23. 9   **24.** 8
25. 8   **26.** 1
27. 0   **28.** 7
29. 10   **30.** 9
31. 0   **32.** 5

33.–41. Answers may vary.

33. $9 - 2 = 7$ **34.** $10 - 1 = 9$
35. $7 - 2 = 5$ **36.** $4 - 4 = 0$
37. $6 - 2 = 4$ **38.** $16 - 8 = 8$
39. $8 - 7 = 1$ **40.** $10 - 4 = 6$
41. $2 - 0 = 2$

Page 58

1. 9
2. 1

Use addition to check your answer.

$$\begin{array}{r} 8 \\ -\ 7 \\ \hline 1 \end{array} \qquad \begin{array}{r} 1 \\ +\ 7 \\ \hline 8 \end{array}$$

3. 6   **4.** 8
5. 8   **6.** 4
7. 9   **8.** 8
9. 9   **10.** 3
11. 8   **12.** 2
13. 5   **14.** 7
15. 11   **16.** 7
17. 9   **18.** 6

Page 59

1. 11 degrees   **2.** 26 degrees

$$\begin{array}{r} 62 \\ -\ 51 \\ \hline 11 \end{array} \text{ degrees} \qquad \begin{array}{r} 69 \\ -\ 42 \\ \hline 27 \end{array} \text{ degrees}$$

3. 37 degrees
    97
  − 60
  37 degrees

4. 13 degrees
    49
  − 36
  13 degrees

5. 44 degrees
    88
  − 44
  44 degrees

6. 40 degrees
    81
  − 41
  40 degrees

Page 60

1. 44
2. 17

Subtract the ones. Use the subtraction facts. $9 - 2 = 7$. Then subtract the tens. $8 - 7 = 1$.

   89
 − 72
   17

   72
 + 17
   89

3. 21   **4.** 53
5. 43   **6.** 12
7. 21   **8.** 45
9. 28   **10.** 12
11. 67   **12.** 21
13. 50
14. 40

Subtract the ones. $9 - 9 = 0$. Subtract the tens. $7 - 3 = 4$.

   79
 − 39
   40

15. 10   **16.** 5
17. 31   **18.** 32
19. 23   **20.** 40

Page 61

1. 31
2. 73

Subtract the ones. $7 - 4 = 3$. Then subtract the tens. $7 - 0 = 7$.

   77
 − 4
   73

   73
 + 4
   77

3. 93   **4.** 81
5. 12   **6.** 22
7. 54   **8.** 94
9. 21 goldfish   **10.** 42 guppies
   29   46
 − 8   − 4
  21 goldfish   42 guppies

Page 62

1. 420
2. 115

Subtract the ones. $6 - 1 = 5$. Then subtract the tens. $7 - 6 = 1$. Then subtract the

hundreds. $8 - 7 = 1$.

  876
 − 761
  115

  115
 + 761
  876

3. 1,165   **4.** 201
5. 100   **6.** 435
7. 7,060   **8.** 5,871
9. 2,121
10. 4,663

Line up the digits. Subtract the ones, tens, and hundreds.

  4,899
 − 236
  4,663

  4,663
 + 236
  4,899

11. 2,712   **12.** 56 feet
                      1,456
                   − 1,400
                56 more feet

13. $321
  $786
 − 465
  $321 more on Sunday

Page 63

1. 3   **2.** 8
3. 8   **4.** 14
5. 17   **6.** 7
7. 11   **8.** 7
9. 15   **10.** 9
11. 13   **12.** 99
13. 3   **14.** 94
15. 20   **16.** 49
17. 23   **18.** 60
19. 34   **20.** 146
21. 433   **22.** 8,999
23. 1,984   **24.** 4,431
25. 2,899   **26.** 5,125
27. $30   **28.** $10
 $15 + $15 = $30 $40 − $30 = $10

Page 65

1. Step 1.
$1,850 can save
$730 money spent
Step 2. Since you need a difference, subtract.
Step 3. $1,850 − $730
Step 4. $1,850 − $730 = $1,120
Step 5. Carolyn will save $1,120.

2. Step 1.
36 cats in 1989
52 cats in 1990
72 cats in 1991
41 cats in 1992

Step 2. You need to add to find out how many cats were sold all together.
Step 3. 36 + 52 + 72 + 41
Step 4. 36 + 52 + 72 + 41 = 201
Step 5. Della sold 201 cats in four years.

Page 66

1. 18
2. 19

Since you can't subtract 3 from 2, borrow 1 ten. Rename the borrowed ten as 10 ones and add it to the 2 ones. 12 − 3 = 9 ones. Then subtract the tens. 8 − 7 = 1 ten.

$$
\begin{array}{r}
{}^{8}9\,{}^{12}2 \\
-\ 7\,3 \\
\hline
1\,9
\end{array}
\qquad
\begin{array}{r}
1 \\
1\,9 \\
+\ 7\,3 \\
\hline
9\,2
\end{array}
$$

3.	18	**4.**	38
5.	45	**6.**	25
7.	17	**8.**	75
9.	8	**10.**	3
11.	9	**12.**	8
13.	25	**14.**	28
15.	48	**16.**	6
17.	39	**18.**	7
19.	65	**20.**	68

Page 67

1. 18
2. 19

Borrow 1 ten and rename. Subtract the ones 13 − 4 = 9. Subtract the tens. 4 − 3 = 1 ten.

$$
\begin{array}{r}
{}^{4}5\,{}^{13}3 \\
-\ 3\,4 \\
\hline
1\,9
\end{array}
\qquad
\begin{array}{r}
1 \\
1\,9 \\
+\ 3\,4 \\
\hline
5\,3
\end{array}
$$

3.	15	**4.**	34
5.	9	**6.**	13
7.	8	**8.**	4
9.	73	**10.**	59
11.	39	**12.**	29
13.	$17	**14.**	$28

$$
\begin{array}{r}
{}^{3}\$4\,{}^{15}5 \\
-\ 2\,8 \\
\hline
\$1\,7
\end{array}
\qquad
\begin{array}{r}
{}^{6}\$7\,{}^{14}4 \\
-\ 4\,6 \\
\hline
\$2\,8
\end{array}
$$

Page 68

1. 17
2. 24

Borrow 1 ten and rename as 10 ones. Subtract the ones. 10 − 6 = 4 ones. Then subtract the tens. 4 − 2 = 2 tens.

$$
\begin{array}{r}
{}^{4}5\,{}^{10}0 \\
-\ 2\,6 \\
\hline
2\,4
\end{array}
\qquad
\begin{array}{r}
1 \\
2\,4 \\
+\ 2\,6 \\
\hline
5\,0
\end{array}
$$

3.	31	**4.**	26
5.	6	**6.**	7
7.	2	**8.**	9
9.	26	**10.**	43
11.	68	**12.**	81
13.	37	**14.**	75
15.	19	**16.**	4
17.	15	**18.**	62
19.	21	**20.**	41

Page 69

1. 45
2. 23

Rename 1 ten as 10 ones. Subtract the ones. 10 − 7 = 3 ones. Subtract the tens 2 − 0 = 2 tens.

$$
\begin{array}{r}
{}^{2}3\,{}^{10}0 \\
-\ \ 7 \\
\hline
2\,3
\end{array}
\qquad
\begin{array}{r}
2\,3 \\
+\ \ 7 \\
\hline
3\,0
\end{array}
$$

3.	31	**4.**	4
5.	40	**6.**	59
7.	74	**8.**	1
9.	7	**10.**	10
11.	3	**12.**	50
13.	18	**14.**	$11

$$
\begin{array}{r}
{}^{8}9\,{}^{10}0 \\
-\ 7\,2 \\
\hline
\$1\,8
\end{array}
\qquad
\begin{array}{r}
{}^{2}\$3\,{}^{11}0 \\
-\ 1\,9 \\
\hline
\$1\,1
\end{array}
$$

Page 70

1.	8	**2.**	7
3.	14	**4.**	38
5.	31	**6.**	111
7.	160	**8.**	235
9.	2,810	**10.**	1,595
11.	35	**12.**	104
13.	8	**14.**	91
15.	65	**16.**	35
17.	78	**18.**	9
19.	23	**20.**	25
21.	42	**22.**	72
23.	46	**24.**	9
25.	20	**26.**	3
27.	14 hours	**28.**	42 hours

$$
\begin{array}{r}
{}^{2}3\,{}^{10}0 \\
-\ 1\,6 \\
\hline
1\,4\ \text{hours}
\end{array}
\qquad
\begin{array}{r}
2\,2 \\
+\ 2\,0 \\
\hline
4\,2\ \text{hours}
\end{array}
$$

Page 71

1.

	Nails	Screws	Washers	Nuts
Amount needed in stock	80	74	60	45
In stock at end of the month	78	65	52	39
Amount Brenda should order	2	9	8	6

2.

	Monday	Tuesday	Wednesday	Thursday	Friday	Saturday
Beginning Count	87	80	66	54	47	38
Cans Sold	7	14	12	7	9	28
Ending Count	80	66	54	47	38	10

Page 72

1. 177

2. 188

Rename 1 ten as 10 ones. Subtract the ones 11 − 3 = 8 ones. Rename 1 hundred as 10 tens. Subtract the tens. 15 − 7 = 8 tens. Subtract the hundreds. 4 − 3 = 1 hundred.

$$\begin{array}{r} \overset{\;\;\;\;15}{\overset{4\;\;\overset{}{5}\;11}{\cancel{5}\,\cancel{6}\,\cancel{1}}} \\ -\;3\,7\,3 \\ \hline 1\,8\,8 \end{array} \qquad \begin{array}{r} \overset{1\;1}{\;} \\ 1\,8\,8 \\ +\;3\,7\,3 \\ \hline 5\,6\,1 \end{array}$$

3. 759
4. 229
5. 108
6. 28
7. 207
8. 195
9. 84
10. 75
11. 190
12. 176
13. 389
14. 209
15. 299
16. 357

Page 73

1. 1,657

2. 2,907

Rename 1 ten as 10 ones. 14 − 7 = 7 ones. Subtract tens. 8 − 8 = 0 tens. To subtract hundreds, rename 1 thousand as 10 hundreds. 17 − 8 = 9 hundreds. Subtract the thousands. 3 − 1 = 2 thousands.

$$\begin{array}{r} \overset{3\;17\;8\;14}{\cancel{4},\cancel{7}\,\cancel{9}\,\cancel{4}} \\ -\;1,8\,8\,7 \\ \hline 2,9\,0\,7 \end{array} \qquad \begin{array}{r} \overset{1\;\;\;1}{\;} \\ 2,9\,0\,7 \\ +\;1,8\,8\,7 \\ \hline 4,7\,9\,4 \end{array}$$

3. 3,092
4. 7,173
5. 10,908
6. 35,748
7. 75,726
8. 170,612
9. 887,018

Page 74

1. 89
2. 579
3. 1,506
4. 253
5. 183
6. 683
7. 2,340
8. 10,838
9. 12,126
10. 97,334
11. 2,355
12. 6,011
13. 51,373
14. 98,738
15. 950,153
16. 75,169
17. 87,220 miles
18. 855,101 miles

$$\begin{array}{r} 88,640 \\ -\;1,420 \\ \hline 87,220 \;\text{miles} \end{array} \qquad \begin{array}{r} \overset{5\;12\;10}{8\,\cancel{6}\,\cancel{3},\cancel{0}\,2\,7} \\ -\;\;\;\;7,9\,2\,6 \\ \hline 8\,5\,5,1\,0\,1 \;\text{miles} \end{array}$$

Page 75

1. 237

2. 202

To subtract ones, rename. There are no tens. Rename 1 hun-dred as 10 tens. Rename 1 ten as 10 ones. Now you have 2 hundreds, 9 tens, and 10 ones. Subtract. 10 − 8 = 2 ones. 9 − 9 = 0 tens. 2 − 0 = 2 hundreds.

$$\begin{array}{r} \overset{2\;\;9\;10}{\cancel{3}\,\cancel{0}\,\cancel{0}} \\ -\;\;9\,8 \\ \hline 2\,0\,2 \end{array} \qquad \begin{array}{r} \overset{1\,1}{\;} \\ 2\,0\,2 \\ +\;\;9\,8 \\ \hline 3\,0\,0 \end{array}$$

3. 344
4. 502
5. 1,458
6. 3,193
7. 2,007
8. 8,223
9. 844
10. 17,441
11. 53,550
12. 85,478
13. 3,859
14. 21,852
15. 65,994
16. 40,818

Page 76

1. 3,505

2. 297

There are no ones, tens, or hundreds. Rename 1 thousand as 10 hundreds, rename 1 hundred as 10 tens, and rename 1 ten as 10 ones. Now you have 6 thousands, 9 hundreds, 9 tens, and 10 ones. Subtract. 10 − 3 = 7 ones. 9 − 0 = 9 tens. 9 − 7 = 2 hundreds. 6 − 6 = 0 thousands.

$$\begin{array}{r} \overset{9\ 9}{\underset{6\ \cancel{10}\ \cancel{10}\ 10}{\ }} \\ 7{,}\cancel{0}\ \cancel{0}\ \cancel{0} \\ -\ 6{,}7\ 0\ 3 \\ \hline 2\ 9\ 7 \end{array} \qquad \begin{array}{r} 1\ 11 \\ 297 \\ +\ 6{,}703 \\ \hline 7{,}000 \end{array}$$

3. 15   **4.** 2,230
5. 23,109   **6.** 26,593
7. 96,604   **8.** 180,070
9. 5,074   **10.** 18,328
11. 536,341   **12.** 257,381
13. 243,968
There are no ones, tens, hundreds, thousands, or ten thousands. Rename to get 2 hundred thousands, 9 ten thousands, 9 thousands, 9 hundreds, 9 tens, and 10 ones. Subtract. $10 - 2 = 8$ ones. $9 - 3 = 6$ tens. $9 - 0 = 9$ hundreds. $9 - 6 = 3$ thousands. $9 - 5 = 4$ ten thousands. $2 - 0 = 2$ hundred thousands.

$$\begin{array}{r} \overset{9\ 9\ 9\ 9}{\underset{2\ \cancel{10}\ \cancel{10}\ \cancel{10}\ \cancel{10}\ 10}{\ }} \\ 3\ \cancel{0}\ \cancel{0}\ \cancel{0}\ \cancel{0}\ \cancel{0} \\ -\quad 5\ 6{,}0\ 3\ 2 \\ \hline 2\ 4\ 3{,}9\ 6\ 8 \end{array} \qquad \begin{array}{r} 1\ 1\ 1\ 1\ 1 \\ 2\ 4\ 3{,}9\ 6\ 8 \\ +\quad 5\ 6{,}0\ 3\ 2 \\ \hline 3\ 0\ 0{,}0\ 0\ 0 \end{array}$$

14. 663,725

Page 77

1. **449**
2. 102
To subtract ones, rename. There are no tens. Rename 1 hundred as 10 tens. Rename 1 ten as 10 ones. Now you have 2 hundreds, 9 tens, and 11 ones. Subtract $11 - 9 = 2$ ones. $9 - 9 = 0$ tens. $2 - 1 = 1$ hundred.

$$\begin{array}{r} \overset{9}{\underset{2\ \cancel{10}\ 11}{\ }} \\ \cancel{3}\ \cancel{0}\ 1 \\ -\ 1\ 9\ 9 \\ \hline 1\ 0\ 2 \end{array} \qquad \begin{array}{r} 1\ 1 \\ 102 \\ +\ 199 \\ \hline 301 \end{array}$$

3. 139   **4.** 479
5. 129   **6.** 347
7. 415   **8.** 608
9. 1,255   **10.** 3,529
11. 5,369   **12.** 4,706
13. 157   **14.** 208
15. 1,067

Page 78

1. **2,516**
2. 2,109
Rename to get 3 thousands, 9 hundreds, 9 tens, and 11 ones. Subtract. $11 - 2 = 9$ ones. $9 - 9 = 0$ tens. $9 - 8 = 1$

---

hundred. $3 - 1 = 2$ thousands.

$$\begin{array}{r} \overset{9\ 9}{\underset{3\ \cancel{10}\ \cancel{10}\ 11}{\ }} \\ 4{,}\cancel{0}\ \cancel{0}\ \cancel{1} \\ -\ 1{,}8\ 9\ 2 \\ \hline 2{,}1\ 0\ 9 \end{array} \qquad \begin{array}{r} 1\ 11 \\ 2{,}109 \\ +\ 1{,}892 \\ \hline 4{,}001 \end{array}$$

3. 9   **4.** 1,659
5. 6,317   **6.** 10,857
7. 17,351   **8.** 9,267
9. 88,533   **10.** 114,086
11. 2,076   **12.** 18,452
13. 25,407

Page 79

1. 19,000 feet 3 < 5, so 19,340 rounds down to 19,000.
2. 23,000 feet
8 > 5, so 22,831 rounds up to 23,000.

Page 80

3. 14,000 feet 4 < 5, so 14,410 rounds down to 14,000.
4. 14,000 feet
6 > 5, so 13,677 rounds up to 14,000.
5. 12,000 feet
3 < 5, so 12,388 rounds down to 12,000.
6. Kilimanjaro
19,000 > 14,000
7. Mount McKinley   **8.** 2,000 feet
12,000 < 20,000

$$\begin{array}{r} 14{,}000 \\ -\ 12{,}000 \\ \hline 2{,}000 \end{array}$$

9. 9,000 feet   **10.** 9,000 feet

$$\begin{array}{r} 23{,}000 \\ -\ 14{,}000 \\ \hline 9{,}000 \end{array} \qquad \begin{array}{r} 29{,}000 \\ -\ 20{,}000 \\ \hline 9{,}000 \end{array}$$

11. 3,000 feet   **12.** 3,000 feet

$$\begin{array}{r} 20{,}000 \\ -\ 17{,}000 \\ \hline 3{,}000 \end{array} \qquad \begin{array}{r} 19{,}000 \\ -\ 16{,}000 \\ \hline 3{,}000 \end{array}$$

Unit 3 Review, page 81

1. 9   **2.** 6
3. 8   **4.** 14
5. 11   **6.** 70
7. 7   **8.** 2
9. 32   **10.** 62
11. 124   **12.** 910
13. 314   **14.** 105
15. 52   **16.** 140
17. 1,023   **18.** 36
19. 17   **20.** 9
21. 7   **22.** 9
23. 35   **24.** 10

25.	43	26.	5
27.	1	28.	34
29.	47	30.	54
31.	75	32.	14
33.	8		

Page 82

34.	69	35.	207
36.	2,059	37.	1,055
38.	3,696	39.	1,897
40.	16,003	41.	107
42.	46	43.	373
44.	1,538	45.	1,448
46.	62,346	47.	88,943
48.	11,373	49.	206,726
50.	665,099		

Unit 4

Page 83

1. 2 ones
2. 5 hundreds
   The 5 is in the third place to the left, the hundreds place.

3.	6 tens	4.	1 thousand
5.	6 thousands	6.	0 tens

Page 84

7.	353	8.	1,397

8. Line up the digits and add.
```
  1,365
+    32
------
  1,397
```

9.	16,572	10.	345
11.	2,010	12.	103,000

11. Line up the digits and add.
```
  1 11
  1,619
+   391
------
  2,010
```

13.	791	14.	1,712

14. Line up the digits and add.
```
      1
  1,709
+     3
------
  1,712
```

15.	543	16.	15,912
17.	56,313	18.	348
19.	3,121	20.	5,900
21.	32,513	22.	14,020

Page 85

1.	0	1	2	3	4	5	6	7	8	9
2.	0	2	4	6	8	10	12	14	16	18
3.	0	3	6	9	12	15	18	21	24	27
4.	0	4	8	12	16	20	24	28	32	36
5.	0	5	10	15	20	25	30	35	40	45
6.	0	6	12	18	24	30	36	42	48	54
7.	0	7	14	21	28	35	42	49	56	63
8.	0	8	16	24	32	40	48	56	64	72
9.	0	9	18	27	36	45	54	63	72	81

Page 86

×	0	1	2	3	4	5	6	7	8	9
0	0	0	0	0	0	0	0	0	0	0
1	0	1	2	3	4	5	6	7	8	9
2	0	2	4	6	8	10	12	14	16	18
3	0	3	6	9	12	15	18	21	24	27
4	0	4	8	12	16	20	24	28	32	36
5	0	5	10	15	20	25	30	35	40	45
6	0	6	12	18	24	30	36	42	48	54
7	0	7	14	21	28	35	42	49	56	63
8	0	8	16	24	32	40	48	56	64	72
9	0	9	18	27	36	45	54	63	72	81

1. 42
2. 27
   Find the number 3 in farthest column on the left. Move across that row until you reach the column with the number 9 at the top. The number in the box is 27.

3.	35	4.	0
5.	81	6.	32
7.	30	8.	12
9.	56	10.	12
11.	36	12.	6
13.	16	14.	0
15.	5	16.	72
17.	4	18.	21
19.	54	20.	40
21.	8	22.	9
23.	0	24.	10

Page 87

1.	25	2.	16
3.	21	4.	36
5.	8	6.	0
7.	1	8.	5
9.	4	10.	3
11.	6	12.	9
13.	0	14.	45
15.	3	16.	4
17.	5	18.	3
19.	56	20.	30

21.–40. Answers may vary.

21. $5 \times 4 = 20$
$4 \times 5 = 20$

22. $2 \times 8 = 16$
$8 \times 2 = 16$
$4 \times 4 = 16$

23. $3 \times 9 = 27$
$9 \times 3 = 27$

24. $4 \times 8 = 32$
$8 \times 4 = 32$

25. $6 \times 7 = 42$
$7 \times 6 = 42$

26. $2 \times 6 = 12$
$6 \times 2 = 12$
$3 \times 4 = 12$
$4 \times 3 = 12$

27. $1 \times 9 = 9$
$3 \times 3 = 9$
$9 \times 1 = 9$

28. $0 \times 0 = 0$
$5 \times 0 = 0$
$1 \times 0 = 0$
$6 \times 0 = 0$
$2 \times 0 = 0$
$7 \times 0 = 0$
$3 \times 0 = 0$
$8 \times 0 = 0$
$4 \times 0 = 0$
$9 \times 0 = 0$

29. $1 \times 8 = 8$
$8 \times 1 = 8$
$2 \times 4 = 8$
$4 \times 2 = 8$

30. $6 \times 9 = 54$
$9 \times 6 = 54$

31. $5 \times 9 = 45$
$9 \times 5 = 45$

32. $8 \times 8 = 64$

33. $3 \times 5 = 15$
$5 \times 3 = 15$

34. $3 \times 7 = 21$
$7 \times 3 = 21$

35. $4 \times 9 = 36$
$9 \times 4 = 36$
$6 \times 6 = 36$

36. $8 \times 9 = 72$
$9 \times 8 = 72$

37. $3 \times 8 = 24$
$8 \times 3 = 24$
$4 \times 6 = 24$
$6 \times 4 = 24$

38. $5 \times 5 = 25$

39. $2 \times 5 = 10$
$5 \times 2 = 10$

40. $2 \times 9 = 18$
$9 \times 2 = 18$
$3 \times 6 = 18$
$6 \times 3 = 18$

Page 88

1. b
6 boxes of pens at \$2 per box = $6 \times \$2 = \$12$

2. a
$\$12 < \$20$

3. b
$9 \times 6 = 54$

4. c

5. b
$8 \times \$2 = \$16$

6. a
$5 \times \$3 = 15$
$\begin{array}{r} 1 \\ \$15 \\ + \ 16 \\ \hline \$31 \end{array}$
$\$31 < \$40$

Page 89

1. 69

2. 22
Line up the digits. Multiply the ones.
$2 \times 1 = 2$ ones. Multiply the tens.
$2 \times 1 = 2$ tens.
$\begin{array}{r} 11 \\ \times \ 2 \\ \hline 22 \end{array}$

3. 39   **4.** 48
5. 63   **6.** 33
7. 32   **8.** 82
9. 84   **10.** 84
11. 66   **12.** 88
13. 153   **14.** 486
15. 699   **16.** 864
17. 936   **18.** 844
19. 369   **20.** 462
21. 562   **22.** 806
23. 990   **24.** 480
25. 36

26. 69
Line up the digits. Multiply the ones.
$3 \times 3 = 9$ ones. Multiply the tens.
$3 \times 2 = 6$ tens.
$\begin{array}{r} 23 \\ \times \ 3 \\ \hline 69 \end{array}$

27. 408   **28.** 688

Page 90

1. 106

2. 123
Multiply the ones. $3 \times 1 = 3$ ones. Multiply the tens. $3 \times 4 = 12$ ones. Since the answer is more than 10, put the 2 in the tens column and the 1 in the hundreds column.
$\begin{array}{r} 41 \\ \times \ 3 \\ \hline 123 \end{array}$

3. 128   **4.** 1,055
5. 1,648   **6.** 1,266
7. 2,177   **8.** 2,088
9. 21,336   **10.** 12,226
11. 248   **12.** 249
13. 497   **14.** 3,248

15. 156 cookies
$\begin{array}{r} 52 \\ \times \ 3 \\ \hline 156 \ \text{cookies} \end{array}$

16. \$105
$\begin{array}{r} 21 \\ \times \ 5 \\ \hline \$105 \end{array}$

Page 91

1. 1,509

2. 240

Multiply the ones. 4 × 0 = 0 ones. Multiply the tens. 4 × 6 = 24 tens. Put the 4 in the tens column and the 2 in the hundreds column.

$$\begin{array}{r} 60 \\ \times\ 4 \\ \hline 240 \end{array}$$

3. 1,809   **4.** 120
5. 1,408   **6.** 12,008
7. 10,046   **8.** 56,008
9. 18,060   **10.** 16,008
11. 14,007   **12.** 18,039
13. 20,408   **14.** 48,080
15. $1,206

Since 2 weeks × 2 = 4 weeks, multiply $603 × 2.

$$\begin{array}{r} \$603 \\ \times\ 2 \\ \hline \$1,206 \end{array}$$

16. $906

$$\begin{array}{r} \$302 \\ \times\ 3 \\ \hline \$906 \end{array}$$

Page 92

1. 12   **2.** 42
3. 32   **4.** 3
5. 9   **6.** 17
7. 30   **8.** 41
9. 69   **10.** 84
11. 19   **12.** 93
13. 50   **14.** 99
15. 80   **16.** 19
17. 48   **18.** 61
19. 60   **20.** 80
21. 99   **22.** 90
23. 840   **24.** 866
25. 537   **26.** 1,980
27. 6,306   **28.** 6,409
29. 150   **30.** 357
31. 100   **32.** 3,099
33. 18,009   **34.** 1,408
35. 159   **36.** 91
37. 169   **38.** 2,800
39. 24,800   **40.** 9,910
41. 8,575   **42.** 63,000

Page 93

1. 220 pounds   **2.** 800 pounds

Page 94

3. 400 pounds   **4.** 800 pounds

$$\begin{array}{r} 400\ \text{pounds} \\ \times\ 2 \\ \hline 800\ \text{pounds} \end{array}$$

5.
$$\begin{array}{r} 1,600\ \text{pounds} \\ 400\ \text{pounds} \\ \times\ 4 \\ \hline 1,600\ \text{pounds} \end{array}$$

6.
$$\begin{array}{r} 2,400\ \text{pounds} \\ 800\ \text{pounds} \\ \times\ 3 \\ \hline 2,400\ \text{pounds} \end{array}$$

7.
$$\begin{array}{r} 6,400\ \text{pounds} \\ 800\ \text{pounds} \\ \times\ 8 \\ \hline 6,400\ \text{pounds} \end{array}$$

8.
$$\begin{array}{r} 880\ \text{pounds} \\ 220\ \text{pounds} \\ \times\ 4 \\ \hline 880\ \text{pounds} \end{array}$$

9.
$$\begin{array}{r} 1,200\ \text{pounds} \\ 400\ \text{pounds} \\ +\ 800\ \text{pounds} \\ \hline 1,200\ \text{pounds} \end{array}$$

10.
$$\begin{array}{r} 4,800\ \text{pounds} \\ 1,200\ \text{pounds} \\ \times\ 4\ \text{pounds} \\ \hline 4,800\ \text{pounds} \end{array}$$

11.
$$\begin{array}{r} 1,020\ \text{pounds} \\ 220\ \text{pounds} \\ +\ 800\ \text{pounds} \\ \hline 1,020\ \text{pounds} \end{array}$$

12.
$$\begin{array}{r} 4,080\ \text{pounds} \\ 1,020\ \text{pounds} \\ \times\ 4 \\ \hline 4,080\ \text{pounds} \end{array}$$

Page 95

1. 682
2. 408

Multiply by 4 ones. 4 × 2 = 8 ones. 4 × 1 = 4 tens. Write 48. Multiply by 3 tens. 3 × 2 = 6 tens. 3 × 1 = 3 hundreds. Write 36. Add the partial products.

$$\begin{array}{r} 12 \\ \times\ 34 \\ \hline 48 \\ +\ 36 \\ \hline 408 \end{array}$$

3. 299   **4.** 616
5. 504   **6.** 483
7. 168   **8.** 924
9. 890   **10.** 516
11. 990   **12.** 840
13. 750   **14.** 880
15. 660

Page 96

1. 1,092
2. 949

Multiply by 3 ones. 3 × 3 = 9 ones. 3 × 7 = 21 tens. Write 219. Multiply by 1 ten. 1 × 3 = 3 tens. 1 × 7 = 7 hundreds. Write 73. Add the partial products.

$$\begin{array}{r} 73 \\ \times\ 13 \\ \hline 219 \\ +\ 73 \\ \hline 949 \end{array}$$

3. 1,400   **4.** 3,165
5. 21,600   **6.** 25,886
7. 10,880   **8.** 50,400
9. 19,866   **10.** 170,714
11. 137,148   **12.** 1,333,031
13. 965,052   **14.** 132,528

15. 2,240,704   **16.** 4,947,000
17. 1,986,622

Page 97

1. 97,344
2. 48,160
Multiply by 2 ones. $2 \times 0 = 0$ ones. $2 \times 3 = 6$ tens. $2 \times 4 = 8$ hundreds Write 860. Multiply by 1 ten. $1 \times 0 = 0$ tens. $1 \times 3 = 3$ hundreds. $1 \times 4 = 4$ thousands. Write 4,300. Multiply by 1 hundred. $1 \times 0 = 0$ hundreds. $1 \times 3 = 3$ thousands. $1 \times 4 = 4$ ten thousands. Write 43,000. Add the partial products.

$$\begin{array}{r} 430 \\ \times\ 112 \\ \hline 860 \\ 4\ 30 \\ +\ 43\ 0 \\ \hline 48{,}160 \end{array}$$

3. 26,866   **4.** 66,521
5. 85,012   **6.** 29,040
7. 70,503   **8.** 43,400
9. 66,822   **10.** 84,400
11. 97,546   **12.** 82,600

Page 98

1. 122,430
2. 195,000
Multiply by 5 ones. $5 \times 0 = 0$ ones. $5 \times 0 = 0$ tens. $5 \times 6 = 30$ hundreds. Write 3,000. Multiply by 2 tens. $2 \times 0 = 0$ tens. $2 \times 0 = 0$ hundreds. $2 \times 6 = 12$ thousands. Write 12,000. Multiply by 3 hundreds. $3 \times 0 = 0$ hundreds. $3 \times 0 = 0$ thousands. $3 \times 6 = 18$ ten thousands. Write 180,000. Add the partial products.

$$\begin{array}{r} 600 \\ \times\ 325 \\ \hline 3\ 000 \\ 12\ 00 \\ +\ 180\ 0 \\ \hline 196{,}000 \end{array}$$

3. 511,128   **4.** 68,523,000
5. 113,364   **6.** 2,574,800
7. 22,917,129   **8.** 55,436,120
9. 12,432 pages

$$\begin{array}{r} 112\ \text{pages} \\ \times\ 111 \\ \hline 112 \\ 1\ 12 \\ +\ 11\ 2 \\ \hline 12{,}432\ \text{pages} \end{array}$$

10. $224

$$\begin{array}{r} \$112 \\ \times\ \ 2 \\ \hline \$\ 224 \end{array}$$

Page 99

1. 75,025 boxes

$$\begin{array}{r} 3{,}001\ \text{cases} \\ \times\ \ \ 25\ \text{boxes} \\ \hline 15\ 005 \\ +\ 60\ 02 \\ \hline 75{,}025\ \text{boxes} \end{array}$$

2. 180,450 packages

$$\begin{array}{r} 4{,}010\ \text{cases} \\ \times\ \ \ 45\ \text{packages of} \\ \hline 20\ 050\ \text{paper plates} \\ +\ 160\ 40 \\ \hline 180{,}450\ \text{packages} \end{array}$$

3. 360 rolls

$$\begin{array}{r} 10\ \text{shelves} \\ \times\ 36\ \text{rolls} \\ \hline 60 \\ +\ 30 \\ \hline 360\ \text{rolls} \end{array}$$

4. 1,206

$$\begin{array}{r} 402\ \text{boxes} \\ \times\ \ \ 3\ \text{shelves} \\ \hline 1{,}206\ \text{boxes} \end{array}$$

Page 100

1. 420   **2.** 13
3. 48   **4.** 1
5. 660   **6.** 41
7. 389   **8.** 488
9. 1,284   **10.** 21,229
11. 466   **12.** 610
13. 1,134   **14.** 127,920
15. 10,000   **16.** 6,900
17. 39,249   **18.** 28,800
19. 13,000   **20.** 500
21. 20,408   **22.** 300
23. 7,000   **24.** 68,000
25. 307,622   **26.** 1,580
27. 114   **28.** 1,532,599

Page 101

1. 960
2. 5,200
Multiply by 0 ones. Write the zero in the ones column. Multiply by 1. Write the answer to the left of the zero. The answer is the same as the top number, 520, plus 1 zero.

$$\begin{array}{r} 520 \\ \times\ \ 10 \\ \hline 5{,}200 \end{array}$$

3. 10,850   **4.** 78,310
5. 763,000   **6.** 46,100
7. 98,200
Multiply by 0 ones. Write the zero in the ones column. Multiply by 0 tens. Write the 0 in the tens column. Multiply by 1 hundred. Write the answer to the left of the 2 zeros. The answer is the same as the top number, 982, plus 2 zeros.

$$\begin{array}{r} 982 \\ \times\ 100 \\ \hline 98{,}200 \end{array}$$

8. 330,500

9. 4,672,000   **10.**   3,990,000

11. 3,382,000

12. 1,590,000

Multiply by 0 ones. Write the zero in the ones column. Multiply by 0 tens. Write the 0 in the tens column. Multiply by 0 hundreds. Write the 0 in the hundreds column. Multiply by 1 thousand. Write the answer to the left of the 3 zeros. The answer is the same as the top number, 1,590, plus 3 zeros.

$$\begin{array}{r} 1,590 \\ \times\ 1,000 \\ \hline 1,590,000 \end{array}$$

13. 2,706,000   **14.**   25,400,000

15. 17,000,000   **16.**   5,460

17. 23,890   **18.**   47,700

19. 1,290,300   **20.**   2,900

21. 36,000   **22.**   1,109,000

23. 20,000,000

Page 102

1. 270

2. 1,950

Write the number you started with, 195. Since there is 1 zero in 10, put 0 after 195.

3. 34,020   **4.**   **3,100**

5. 28,600

Write the number you started with, 286. Since there are 2 zeros in 100, put 2 zeros after 286.

6. 1,502,900

7. 490,000

8. 1,830,000

Write the number you started with, 1,830. Since there are 3 zeros in 1,000, put 3 zeros after 1,830.

9. 27,600,000   **10.**   2,101,000

11. 13,300   **12.**   69,970

13. 410   **14.**   20,100

15. 33,032,000   **16.**   2,000

17. 310,000   **18.**   21,000,000

19. 890   **20.**   4,200

21. 51,000   **22.**   36,000

23. 500,000   **24.**   9,900

Page 103

1. 600 million bags of potatoes

200 million × 3 = 600 million bags of potatoes

2. 800 million bags of potatoes

200 million × 4 = 800 million bags of potatoes

3. 200 million bags of potatoes

200 million × 1 = 200 million bags of potatoes

4. 800 million bags of potatoes

400 million × 2 years = 800 million bags of potatoes.

Page 104

1. 120 million tons

20 million × 6 = 120 million tons

2. 40 million tons

20 million × 2 = 40 million tons

3. 60 million tons

20 million × 3 = 60 million tons

4. 40,000 million pounds

20 million × 2,000 = 40,000 million pounds

5. 120 million tons

60 million × 2 years = 120 million tons

6. 600 million tons

120 million × 5 years = 600 million tons

Page 105

1. 141

2. 256

Multiply the 2 by 8 ones. 8 × 2 = 16 ones. Rename 16 ones as 1 ten and 6 ones. Write the 6, and carry the 1 ten. Multiply the 3 by 8 ones. 8 × 3 = 24 tens. Add the carried 1 ten. 24 + 1 = 25 tens. Write 25.

$$\begin{array}{r} 1 \\ 32 \\ \times\ 8 \\ \hline 256 \end{array}$$

3. 651   **4.**   180

5. 702   **6.**   324

7. 310   **8.**   282

9. 399   **10.**   738

11. 1,000   **12.**   1,050

13. 1,700   **14.**   2,500

15. 7,875   **16.**   6,412

17. 3,456   **18.**   2,600

19. 2,697   **20.**   1,540

Page 106

1. 48,744

2. 43,968

Multiply by 6 ones. Add the carried numbers.

$$\begin{array}{r} 1\ 1\ 4 \\ 7,3\ 2\ 8 \\ \times6 \\ \hline 4\ 3,9\ 6\ 8 \end{array}$$

3. 17,860   **4.**   27,810

5. 52,694   **6.**   260,900

7.	275,800	**8.**	828,000
9.	13,448	**10.**	95,200
11.	153,086	**12.**	41,934
13.	$73,500	**14.**	$51,400

```
      $24,500
    ×       3
      $73,500
```

```
      $24,500
    +   1,200
      $25,700
          1
      $25,700
    ×       2
      $51,400
```

Page 107

1. **3,054**

2. 1,535

Multiply by 5 ones. Add the carried numbers.

```
        3
      307
    ×   5
    1,535
```

3.	7,248	**4.**	1,435
5.	1,616	**6.**	6,318
7.	2,721	**8.**	3,045
9.	3,036	**10.**	2,416
11.	2,812	**12.**	1,872
13.	6,440	**14.**	3,025
15.	6,464	**16.**	5,418
17.	4,949	**18.**	2,781
19.	1,008	**20.**	1,803
21.	3,236	**22.**	872

Page 108

1. **28,242**

2. 27,219

Multiply by 3 ones. Add the carried numbers.

```
        2
    9,073
    ×   3
    27,219
```

3.	40,445	**4.**	25,235
5.	85,616	**6.**	264,380
7.	160,192	**8.**	818,181
9.	210,040	**10.**	213,542
11.	180,045	**12.**	$2,430

```
          3
      $405
    ×     6
    $2,430
```

13. $24,300

```
    $2,430
    ×    10
    $24,300
```

Page 109

1.	42	**2.**	20
3.	72	**4.**	393
5.	259	**6.**	590
7.	1,206	**8.**	472
9.	936	**10.**	36,064
11.	630,045	**12.**	100
13.	496	**14.**	65
15.	90	**16.**	86
17.	37	**18.**	744
19.	140	**20.**	44
21.	616	**22.**	1,239
23.	2	**24.**	510
25.	8,336	**26.**	693
27.	86,940	**28.**	5,790
29.	388	**30.**	342,664
31.	90	**32.**	1,290,499
33.	1,800	**34.**	2,200
35.	305,624	**36.**	1,104
37.	503,200	**38.**	1,482,103

Page 110

1. **1,610**

2. 1,416

Multiply by 4 ones. $4 \times 9 = 36$ ones. Write 6, and carry the 3. $4 \times 5 = 20$ tens. Add 3 tens. Write 23. Multiply by 2 tens. $2 \times 9 = 18$ tens. Write 8, and carry the 1. $2 \times 5 = 10$ hundreds. Add the carried 1. Write 11. Add the partial products.

```
      59
    × 24
     236
  + 1 18
    1,416
```

3.	3,975	**4.**	3,038
5.	6,450	**6.**	13,338
7.	47,034	**8.**	22,386
9.	27,812	**10.**	17,825
11.	79,398	**12.**	27,500
13.	20,502	**14.**	14,616
15.	6,831		

Page 111

1. **132,900**

2. 73,242

Line up the digits. Multiply by 8 ones. Multiply by 1 ten. Carry when needed. Add the partial products.

```
     4,069
    ×    18
    32 552
  + 40 69
    73,242
```

3. 414,360   **4.** 1,082,664

5. 888,185   **6.** 7,446,231

7. 1,120,590 miles   **8.** 2,388,600 miles

```
    24,902  miles              477,720  miles
×        45                 ×         5
   124 510                  2,388,600  miles
 + 996 08
 1,120,590  miles
```

Page 112

1. 130,410

2. 64,152

Multiply by 2 ones. Multiply by 3 tens.
Multiply by 1 hundred. Carry when
needed. Add the partial products.

```
     486
  × 132
     972
   14 58
 + 48 6
  64,152
```

3. 488,670   **4.** 292,274

5. 338,469   **6.** 366,014

7. 2,684,478   **8.** 11,966,589

9. 14,907,298   **10.** 29,522,932

11. 83,250   **12.** 528,768

13. 544,260   **14.** 13,288,582

Page 113

1. 218,592

2. 123,344

Multiply by 3 ones. Multiply by 9 tens.
Multiply by 5 hundreds. Carry when
needed. Add the partial products.

```
     208
  × 593
     624
   18 72
 + 104 0
  123,344
```

3. 670,714   **4.** 51,435

5. 84,258   **6.** 803,926

7. 3,189,030   **8.** 4,800,474

9. 837,753   **10.** 1,978,880

11. 1,499,300   **12.** 15,554,354

13. 12,853,162   **14.** 30,592,185

15. 38,563,200

Page 114

1. 175,821

2. 387,200

Multiply by 5 ones. Multiply by 0 tens. Put
a 0 in the tens column. Multiply by 6
hundreds, and put the partial product to the
left of the zero. Add the partial products.

```
      640
   × 605
    3 200
    0 00
 + 384 0
  387,200
```

3. 439,488   **4.** 233,640

5. 122,800   **6.** 2,736,000

7. 2,920,000   **8.** 1,706,400

9. 981,400   **10.** 5,400,000

11. 6,512,500   **12.** 45,513,816

13. 29,016,000   **14.** 5,023,012

Page 115

1. 189,958

2. 291,500

Multiply by 0 ones. Write one zero in the
ones column. Multiply by 0 tens. Write one
0 in the tens column. Multiply by 5
hundreds. $5 \times 3 = 15$ hundreds. $5 \times 8 = 40$
thousands. $5 \times 5 = 25$ ten thousands. Write
291,500 to the left of the two zeros.

```
     583
  × 500
  291,500
```

3. 67,946   **4.** 348,870

5. 281,800   **6.** 3,987,456

7. 6,644,800   **8.** 8,344,400

9. 4,396,098   **10.** 16,296,000

11. 62,747,115   **12.** 6,800,000

Page 116

1.
```
 $20,000
   $400
 ×   50
 $20,000
```
2.
```
  $4,000
   $400
 ×   10
  $4,000
```

3.
```
  $1,000
   $100
 ×   10
  $1,000
```
4.
```
   $700
    $70
 ×   10
   $700
```

Page 117

5.
```
   $500
    $50
 ×  10
   $500
```
6.
```
   $500
    $10
 ×  50
   $500
```

7.
```
    $40
    $10
 ×   4
    $40
```
8.
```
  $2,500
    $50
 ×  50
  $2,500
```

9.
```
  $1,000
   $100
 ×  10
  $1,000
```
10. No.
```
    $20
 ×  20
   $400
 $400 < $600
```

Unit 4 Review, page 118

1.	81	2.	35
3.	72	4.	24
5.	24	6.	21
7.	40	8.	0
9.	46	10.	80
11.	328	12.	48
13.	189	14.	663
15.	2,877	16.	426
17.	180	18.	2,109
19.	2,010	20.	63,009
21.	24,800	22.	6,028
23.	504	24.	680
25.	5,061	26.	4,228
27.	19,500	28.	387,903
29.	274,296	30.	96,960
31.	745,984	32.	4,480,056

Page 119

33.	41,480	34.	56,700
35.	209,916	36.	2,962,862
37.	46,299,000	38.	470
39.	1,890	40.	15,700
41.	289,000	42.	2,586,000
43.	10,000,000	44.	3,600
45.	295,000	46.	250
47.	340,600	48.	267
49.	2,043	50.	44,940
51.	46,600	52.	6,814
53.	90,054	54.	320,616
55.	42,140	56.	138,402
57.	891,810	58.	1,875
59.	1,794	60.	24,752
61.	26,708	62.	343,246
63.	362,970		

Page 120

64.	321,245	65.	842,022
66.	1,645,494	67.	1,786,173
68.	560,700	69.	1,166,388
70.	13,383,500	71.	403,088
72.	749,997	73.	105,763
74.	99,770	75.	607,221
76.	79,184	77.	81,200
78.	737,836	79.	8,112,936
80.	19,734,000		

Unit 5

Page 121

1. 10
2. 30

   7 > 5, so 27 rounds up to 30.
3. 50   4. 70

5. 200
6. 600

   5 = 5, so 550 rounds up to 600.
7. 600   8. 400

Page 122

9. 2   10. 114

Line up the digits.
Subtract.

```
   468
 − 354
   114
```

11. 0   12. 109
13. 103   14. 3,343

```
       9 9
     4 10 10 10
     5,0 0 0
   − 1,6 5 7
     3,3 4 3
```

15.	5,065	16.	126
17.	5	18.	8
19.	7	20.	8
21.	192	22.	3,243

```
        47
      × 69
       423
    + 2 82
     3,243
```

23.	28,250	24.	35,224
25.	8	26.	3
27.	9	28.	8
29.	29,385	30.	3,626
31.	575,946	32.	156,624

Page 123

1.	0	1	2	3	4	5	6	7	8	9
2.	0	1	2	3	4	5	6	7	8	9
3.	0	1	2	3	4	5	6	7	8	9
4.	0	1	2	3	4	5	6	7	8	9
5.	0	1	2	3	4	5	6	7	8	9
6.	0	1	2	3	4	5	6	7	8	9
7.	0	1	2	3	4	5	6	7	8	9
8.	0	1	2	3	4	5	6	7	8	9
9.	0	1	2	3	4	5	6	7	8	9

Page 124

1. 4
2. 2

Find the smaller number, 7, in the farthest row on the left. Then move to the right along the row until you find the larger number, 14. Move to the top of the column to find the answer, 2.

3. 4   4. 3
5. 5   6. 6

7.	2	**8.**	7
9.	7	**10.**	5
11.	9	**12.**	9
13.	8	**14.**	6
15.	8	**16.**	7
17.	3	**18.**	8

Page 125

1. 2

2. 3

Find the smaller number, 4 in the farthest row on the left. Then move to the right along the row until you find the larger number, 12. Move to the top of the column to findthe answer, 3.

3.	0	**4.**	2
5.	5	**6.**	6
7.	6	**8.**	9
9.	35	**10.**	42
11.	45	**12.**	0
13.	9	**14.**	9
15.	8	**16.**	5
17.	14	**18.**	4
19.	21	**20.**	5
21.	9	**22.**	6
23.	4	**24.**	36
25.	40	**26.**	7
27.	5	**28.**	6
29.	8	**30.**	64
31.	2	**32.**	7

33.–41. Answers may vary.

33.	$6 \div 2 = 3$	**34.**	$2 \div 1 = 2$
	$3 \div 1 = 3$		$4 \div 2 = 2$
	$9 \div 3 = 3$		$6 \div 3 = 2$
	$12 \div 4 = 3$		$8 \div 4 = 2$
	$15 \div 5 = 3$		$10 \div 5 = 2$
	$18 \div 6 = 3$		$12 \div 6 = 2$
	$21 \div 7 = 3$		$14 \div 7 = 2$
	$24 \div 8 = 3$		$16 \div 8 = 2$
	$27 \div 9 = 3$		$18 \div 9 = 2$
35.	$4 \div 1 = 4$	**36.**	$0 \div 0 = 0$ 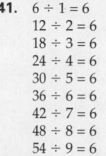
	$8 \div 2 = 4$		$0 \div 1 = 0$
	$12 \div 3 = 4$		$0 \div 2 = 0$
	$16 \div 4 = 4$		$0 \div 3 = 0$
	$20 \div 5 = 4$		$0 \div 4 = 0$
	$24 \div 6 = 4$		$0 \div 5 = 0$
	$28 \div 7 = 4$		$0 \div 6 = 0$
	$32 \div 8 = 4$		$0 \div 7 = 0$
	$36 \div 9 = 4$		$0 \div 8 = 0$
			$0 \div 9 = 0$
37.	$8 \div 1 = 8$	**38.**	$7 \div 1 = 7$
	$16 \div 2 = 8$		$14 \div 2 = 7$
	$24 \div 3 = 8$		$21 \div 3 = 7$

$32 \div 4 = 8$		$28 \div 4 = 7$
$40 \div 5 = 8$		$35 \div 5 = 7$
$48 \div 6 = 8$		$42 \div 6 = 7$
$56 \div 7 = 8$		$49 \div 7 = 7$
$64 \div 8 = 8$		$56 \div 8 = 7$
$72 \div 9 = 8$		$63 \div 9 = 7$

39.	$5 \div 1 = 5$	**40.**	$9 \div 1 = 9$
	$10 \div 2 = 5$		$18 \div 2 = 9$
	$15 \div 3 = 5$		$27 \div 3 = 9$
	$20 \div 4 = 5$		$36 \div 4 = 9$
	$25 \div 5 = 5$		$45 \div 5 = 9$
	$30 \div 6 = 5$		$54 \div 6 = 9$
	$35 \div 7 = 5$		$63 \div 7 = 9$
	$40 \div 8 = 5$		$72 \div 8 = 9$
	$45 \div 9 = 5$		$81 \div 9 = 9$
41.	$6 \div 1 = 6$		
	$12 \div 2 = 6$		
	$18 \div 3 = 6$		
	$24 \div 4 = 6$		
	$30 \div 5 = 6$		
	$36 \div 6 = 6$		
	$42 \div 7 = 6$		
	$48 \div 8 = 6$		
	$54 \div 9 = 6$		

Page 126

1. 5

2. 6

Check division by multiplying the answer, 6, by the number you divided by, 6.

$$\begin{array}{r} 6 \\ \times\,6 \\ \hline 36 \end{array}$$

3.	9	**4.**	9
5.	5	**6.**	7
7.	9	**8.**	8
9.	0	**10.**	7
11.	7	**12.**	8
13.	5	**14.**	7
15.	8	**16.**	9

Page 127

1. 4

2. 9

Divide. Since you can't divide 4 by 5 evenly, divide 45 by 5. $45 \div 5 = 9$. Check division by multiplying the answer, 9, by the number you divided by, 5.

$$5\overline{)45}^{\,9} \qquad \begin{array}{r} 9 \\ \times\,5 \\ \hline 45 \end{array}$$

3.	9	**4.**	6
5.	21	**6.**	32

195

7.	72	8.	61
9.	21	10.	21
11.	71	12.	51
13.	93	14.	42
15.	63	16.	71
17.	91	18.	71
19.	61	20.	64

Page 128

1. 42
2. 64

Divide. Since you can't divide 3 by 6 evenly, divide 38 by 6. $38 \div 6$ isn't a basic fact, so use the closest fact, $36 \div 6 = 6$. Multiply. $6 \times 6 = 36$. Subtract. $38 - 36 = 2$. Bring down the 4. $24 \div 6 = 4$. Check by multiplying.

```
      64
6 ) 384          64
  − 36          ×  6
     24         384
   − 24
      0
```

3.	55	4.	46
5.	82	6.	94
7.	86	8.	54
9.	86	10.	95
11.	56	12.	48
13.	32	14.	53
15.	67	16.	52

Page 129

1. 5 R2
2. 3 R1

Use the basic facts. $10 \div 3 = 3$, with an amount left over. Multiply $3 \times 3 = 9$. Subtract. $10 - 9 = 1$. The remainder is 1. Check by multiplying and adding the remainder.

```
     3  R1
3 ) 10           3
   − 9          × 3
     1           9
               + 1
                10
```

3.	3 R3	4.	8 R3
5.	3 R2	6.	5 R4
7.	5 R1	8.	8 R2
9.	5 R1	10.	7 R1
11.	9 R7	12.	7 R1
13.	7 R3	14.	8 R1
15.	5 R4	16.	5 R2

Page 130

1. 29 R5
2. 38 R2

Divide. $19 \div 5 = 3$, with an amount left over. Multiply. $3 \times 5 = 15$. Subtract. $19 - 15 = 4$. Bring down 2. Divide. $42 \div 5 = 8$ and an amount left over. Multiply. $5 \times 8 = 40$. Subtract. $42 - 40 = 2$. 2 is the remainder. Check by multiplying and adding the remainder.

```
      38  R2
5 ) 192          38
  − 15          ×  5
     42         190
   − 40         +  2
      2         192
```

3.	45 R2	4.	47 R1
5.	63 R2	6.	98 R3
7.	82 R1	8.	72 R4
9.	77 R2	10.	59 R1
11.	69 R1	12.	94 R2

Page 131

1. 214 R1
2. 224 R2

Divide. $6 \div 3 = 2$. Multiply. $2 \times 3 = 6$. Subtract. $6 - 6 = 0$. Bring down the 7. $7 \div 3 = 2$, plus an amount left over. Multiply. $2 \times 3 = 6$. Subtract. $7 - 6 = 1$. Bring down the 4. $14 \div 3 = 4$, plus an amount left over. Multiply. $4 \times 3 = 12$. Subtract. $14 - 12 = 2$. The remainder is 2.

```
      224  R2
3 ) 674          224
  − 6           ×  3
    07          672
   − 6          +  2
    14          674
  − 12
     2
```

3.	239 R1	4.	218 R3
5.	2,371 R2	6.	2,663 R2
7.	1,131 R3	8.	1,122 R3
9.	1,337 R2	10.	1,347 R3
11.	1,763 R1	12.	1,294 R5

Page 132

1. $36

Divide the total, $144, by 4.

```
     $ 36
4 ) $144
  − 12
     24
   − 24
      0
```

Divide. 14 ÷ 4 = 3, plus an amount left over. Multiply. 3 × 4 = 12. Subtract. 14 − 12 = 2. Bring down the 4. Divide. 21 ÷ 4 = 6. Multiply. 6 × 4 = 24. Subtract. 24 − 24 = 0. There is no remainder. Each person will have to pay $36.

2. $6

```
      $ 6
  4 )$24
    − 24
       0
```

3. $145

```
      $145
  4 )$580
    − 4
      18
    − 16
      20
    − 20
       0
```

4. $116

```
      $116
  5 )$580
    − 5
      08
    − 5
      30
    − 30
       0
```

Page 133

1.	8	2.	30
3.	9	4.	8
5.	7	6.	17
7.	4	8.	8
9.	41	10.	128
11.	72	12.	69
13.	609	14.	91
15.	350	16.	21
17.	2 R4	18.	7 R2
19.	66	20.	7 R2
21.	250	22.	3 R5
23.	364	24.	8 R1
25.	17	26.	1,980
27.	98	28.	37 R3
29.	662	30.	65
31.	522	32.	556
33.	88 R1		

Page 134

1. 438 R5
2. 269

Divide. 16 ÷ 6 = 2, with an amount left over. Multiply. 2 × 6 = 12. Subtract. 16 − 12 = 4. Bring down the 1. Divide. 41 ÷ 6 = 6, plus an amount left over. Multiply. 6 × 6 = 36. Subtract. 41 − 36 = 5. Bring down the 4. Divide. 54 ÷ 6 = 9. Multiply. 9 × 6 = 54. Subtract. 54 − 54 = 0. There is no remainder.

```
      269            269
  6 )1,614          ×  6
    − 12           1,614
      41
    − 36
      54
    − 54
       0
```

3.	427 R2	4.	644 R2
5.	733 R3	6.	728 R4
7.	243	8.	368 R3
9.	889 R1	10.	469 R3
11.	633 R1	12.	424 R2

Page 135

1. 361 R8
2. 861

Divide. 68 ÷ 8 = 8, plus an amount left over. Multiply. 8 × 8 = 64. Subtract. 68 − 64 = 4. Bring down the 8. Divide. 48 ÷ 8 = 6. Multiply. 6 × 8 = 48. Subtract. 48 − 48 = 0. Bring down the 8. Divide. 8 ÷ 8 = 1. Multiply. 1 × 8 = 8. Subtract. 8 − 8 = 0. There is no remainder.

```
      861            861
  8 )6,888          ×  8
    − 64           6,888
      48
    − 48
      08
    − 8
       0
```

3.	6,122 R5	4.	9,581
5.	6,314	6.	5,295 R1
7.	16,897 R1	8.	7,158 R5
9.	9,215 R1	10.	9,126 R2

Page 136

1. 15
2. 26 R1

Divide. 20 ÷ 8 = 2, plus an amount left over. Multiply. 2 × 8 = 16. Subtract. 20 − 16 = 4. Bring down the 9. Divide. 49 ÷ 8 = 6, plus an amount left over. Multiply 6 × 8 = 48. Subtract 49 − 48 = 1. There is a remainder of 1.

```
      26  R1          26
  8 )209            ×  8
    − 16            208
      49          +   1
    − 48            209
       1
```

3.	84	4.	81 R1
5.	751		

6. 286 R6

Divide. 20 ÷ 7 = 2, plus an amount left over. Multiply. 2 × 7 = 14. Subtract. 20 − 14 = 6. Bring down the 0. Divide. 60 ÷ 7 = 8, plus an amount left over. Multiply. 8 × 7 = 56. Subtract. 60 − 56 = 4. Bring down the 8. Divide. 48 ÷ 7 = 6. Multiply. 6 × 7 = 42. Subtract. The remainder is 6.

```
    286  R6
7)2,008              286
 −14               ×    7
  60               2,002
 −56             +      6
  48               2,008
 −42
   6
```

7. 833 R5   **8.** 1,625 R7
9. 5,358 R2   **10.** 1,667 R4

Page 137

1. 400
2. 600

Divide. 30 ÷ 5 = 6. Multiply, subtract, and bring down 0. 6 × 5 = 30. 30 − 30 = 0. Divide. 0 ÷ 5 = 0. Multiply, subtract, and bring down the last 0. Multiply and subtract. There is no remainder.

```
    600
5)3,000            600
 −30             ×   5
  00             3,000
 − 0
  00
 − 0
   0
```

3. 400   **4.** 2,000
5. 8,000   **6.** 6,000
7. 6,000   **8.** 4,000
9. 8,000   **10.** 9,000
11. 7,000   **12.** 8,000

Page 138

1. $40

Round $209 to the nearest hundred. $209 rounds to $200. To split $200 into 5 equal amounts, divide by 5.

```
   $40
5)$200            $40
 −20            ×  5
  00            $200
 − 0
   0
```

She will pay about $40 each month.

2. $50

```
   $ 50
8)$400            $50
 −40            ×  8
  00            $400
 − 0
   0
```

She will pay about $50 each month.

3. $120

```
   $120
5)$600            $120
 −5             ×   5
 10             $600
−10
 00
 − 0
   0
```

He will pay about $120 each month.

4. $75

```
   $ 75
8)$600            $75
 −56            ×  8
  40            $600
 −40
   0
```

Page 139

5. $2,400

```
    $2,400
2)$4,800          $2,400
 −4             ×     2
 08             $4,800
 −8
 00
 − 0
 00
 − 0
  0
```

Each payment will be about $2,400.

6. $800

```
    $800
6)$4,800          $800
 −48            ×    6
  00            $4,800
 − 0
  00
 − 0
   0
```

He will pay about $800 each month.

7. $1,200

```
      $1,200
   4) $4,800      $1,200
    − 4           ×    4
      0 8         $4,800
    − 8
      00
    − 0
      00
    − 0
      0
```

He will pay about $1,200 each month.

8. $1,025

```
      $1,025
   8) $8,200      $1,025
    − 8           ×      8
      0 20        $8,200
    − 16
      40
    − 40
      0
```

Each payment will be about $1,025.

9. $300

```
      $300
   2) $600        $300
    − 6           ×   2
      00          $600
    − 0
      00
    − 0
      0
```

Each payment will be about $300.

10. $50

```
      $50
   12) $600       $50
    − 60          ×  12
      00          100
    − 0           + 50
      0           $600
```

Each payment will be about $5.

11. $200

```
      $200
   2) $400        $200
    − 4           ×   2
      00          $400
    − 0
      00
    − 0
      0
```

Each person will pay about $200 each month.

12. $50

```
      $50
   4) $200        $50
    − 20          × 4
      00          200
    − 0
      0
```

She will pay about $50 each month.

Page 140

1. 3 R2

2. 4 R2

Estimate how many times 12 goes into 50 by dividing 5 by 1. 5 ÷ 1 = 5. Write 5 above the 0 and multiply. 5 × 12 = 60, which is too large. Try 4. Multiply. 4 × 12 = 48. Subtract. 50 − 48 = 2. The answer is 4, plus a remainder of 2.

```
      4 R2
   12) 50          12
    − 48          ×  4
      2           48
                  + 2
                  50
```

3.	3	**4.**	2
5.	15	**6.**	16 R2
7.	12 R5	**8.**	13 R3
9.	29 R15	**10.**	12
11.	14 R30	**12.**	13 R7

Page 141

1. 3 R24

2. 7 R8

Estimate how many times 22 goes into 162 by dividing 16 by 2. 16 ÷ 2 = 8. Write 8 above the 2. Multiply. 8 × 22 = 176, which is too large. Try 7. 7 × 22 = 154. Subtract. 162 − 154 = 8. The answer is 7, plus a remainder of 8.

```
       7 R8
   22) 162          22
    − 154          ×  7
      8            154
                   +  8
                   162
```

3.	7 R29	**4.**	6 R3
5.	45	**6.**	29 R26
7.	52 R28	**8.**	86 R28
9.	87 R25	**10.**	71 R18
11.	77 R60		

Page 142

1. 169 R18

2. 96

Divide. 51 rounds down to 50. Estimate how many times 51 goes into 4,896 by dividing 48 by 5. 48 ÷ 5 = 9. Put 9 above the 9. Multiply. 9 × 51 = 459. Subtract, and bring down. 489 − 459 = 30. Estimate how many times 51 goes into 306 by dividing 30 by 5. 30 ÷ 5 = 6. Multiply. 6 × 51 = 306. Subtract. 306 − 306 = 0. There is no remainder.

$$
\begin{array}{r}
96 \\
51\overline{)4{,}896} \\
-4\,59 \\
\hline
306 \\
-306 \\
\hline
0
\end{array}
\qquad
\begin{array}{r}
96 \\
\times 51 \\
\hline
96 \\
+480 \\
\hline
4{,}896
\end{array}
$$

3. 51 R56 **4.** 114 R20
5. 1121 **6.** 751 R21
7. 566 R10 **8.** 1159
9. 812 R16 **10.** 1124 R8
11. 1912

Page 143

1. 35
2. 31

Divide, multiply, subtract, and bring down.

$$
\begin{array}{r}
31 \\
30\overline{)930} \\
-90 \\
\hline
30 \\
-30 \\
\hline
0
\end{array}
\qquad
\begin{array}{r}
31 \\
\times 30 \\
\hline
930
\end{array}
$$

3. 84 R25 **4.** 59 R2
5. 65 **6.** 1,821
7. 2,934 **8.** 988
9. 625 R7 **10.** 1,911
11. 8,631 **12.** 852
13. 466 **14.** 1,312 R36

Page 144

1. 82 **2.** 9
3. 98 **4.** 7
5. 2,623 **6.** 12
7. 13 **8.** 3 R1
9. 100 **10.** 9
11. 2,970 **12.** 49
13. 4,900 **14.** 70
15. 130 **16.** 16,290
17. 182 R32 **18.** 304
19. 26 **20.** 562 R9
21. 45,888 **22.** 48
23. 86

Page 145

24. 2,722 **25.** 76 R3
26. 388,800 **27.** 721 R2
28. 684,432 **29.** 2,943
30. 16,413 **31.** 8,327
32. 117 R20 **33.** 4,862,053
34. 1,570 **35.** 5,837
36. 8,878 **37.** 2,376 R8
38. 818,000 **39.** 629
40. 755 R5 **41.** 926,480
42. 66,572

Page 146

1. Bunty
Find the unit price of Diva paper towels. 94 cents ÷ 2 = 47 cents. Compare the prices. 49 cents > 47 cents. Bunty costs more.

2. Bow Wow Chow
80 cents ÷ 4 = 20 cents. 85 ÷ 5 = 17 cents. 20 > 17. Bow Wow Chow costs more.

3. Italia
63 cents ÷ 8 = 7 cents R7. 90 cents ÷ 10 = 9 cents. 9 > 7 R7. Italia costs more.

4. 3 cents
Find how much one pound of onions costs. 57 cents ÷ 3 = 19 cents. 20 cents − 19 cents = 1 cent. You will save 1 cent per pound or 3 cents.

Page 147

1. 13
2. 12

Divide 8,304 by 692 by estimating how many times 6 goes into 8. 8 ÷ 6 = 1, plus an amount left over. Multiply. 692 × 1 = 692. Subtract. 830 − 692 = 138. Bring down the 4. Divide 1384 by 692 by estimating how many times 6 goes into 13. 13 ÷ 6 = 2, plus an amount left over. Multiply. 692 × 2 = 1384. Subtract. 1384 − 1384 = 0. There is no remainder.

$$
\begin{array}{r}
12 \\
692\overline{)8{,}304} \\
-6\,92 \\
\hline
1\,384 \\
-1\,384 \\
\hline
0
\end{array}
\qquad
\begin{array}{r}
692 \\
\times\ \ 12 \\
\hline
1\,384 \\
+6\,92 \\
\hline
8{,}304
\end{array}
$$

3. 14 R5 **4.** 11 R143
5. 18 **6.** 57
7. 92 R5 **8.** 17
9. 7 R6

Page 148

1. 318
2. 911 R30
   Divide, multiply, subtract, and bring down.

```
          911 R30
653) 594,913              911
    − 587 7             × 653
       7 21             2 733
     − 6 53             45 55
        683           + 546 6
      − 653           594,883
         30         +      30
                      594,913
```

3. 68
4. 242 R214
5. 157
6. 867 R49
7. 817 R151
8. 118 R278
9. 479

Page 149

1. 203
2. 1,002
   Divide, multiply, subtract, and bring down.
   7 is larger than zero, so you can't divide.
   Put 0 in the answer, and bring down the 1.
   7 > 1, so put another zero in the answer and
   bring down the 4. Divide. 14 ÷ 7 = 2.
   Multiply. 2 × 7 = 14. Subtract. 14 − 14 = 0.
   There is no remainder.

```
      1,002
7 ) 7,014            1,002
  − 7              ×     7
    0 0             7,014
  −   0
      01
    −  0
       14
     − 14
        0
```

3. 506
4. 806
5. 402
6. 803 R37
7. 506 R3
8. 103
9. 902 R1

Page 150

1. 306 R2
2. 6,002 R5
   Divide, multiply, subtract, and bring down.
   Repeat as many times as necessary to solve
   the problem.

```
      6,002  R5
9 ) 54,023              6,002
  − 54                ×     9
     0 0               54,018
   −  0              +      5
      02               54,023
    −  0
       23
     − 18
        5
```

3. 5,090 R2
4. 808 R1
5. 5,020
6. 900 R14
7. 750 R19
8. 601
9. 100

Page 151

1. 156 stands

```
          156
100) 15,600
   − 10 0
      5 60
    − 5 00
       600
     − 600
         0
```

   She delivered to 156 stands.

2. 19,665 papers

```
      19,665
3 ) 58,995
  − 3
    28
  − 27
     1 9
   − 18
      19
    − 18
       15
     − 15
        0
```

   She delivered 19,665 papers each day.

3. 806 bundles

```
        806
25) 20,150
  − 20 0
     150
   − 150
       0
```

   She delivered 806 bundles.

4. $2

```
        $2
150) 300
   − 300
       0
```

   He charged $2 for each paper.

Unit 5 Review, page 152

1.	3	**2.**	3
3.	8	**4.**	9
5.	6	**6.**	5
7.	5	**8.**	5
9.	4	**10.**	0
11.	12	**12.**	132
13.	111	**14.**	41
15.	71	**16.**	231
17.	241	**18.**	91
19.	29	**20.**	54
21.	233	**22.**	53 R6
23.	39 R3	**24.**	358 R7
25.	5,579 R1	**26.**	125
27.	432	**28.**	15,141
29.	3,341		

Page 153

30.	2	**31.**	3
32.	3	**33.**	2
34.	9	**35.**	26 R5
36.	77 R45	**37.**	211
38.	72 R5	**39.**	442 R20
40.	937	**41.**	65 R50
42.	216 R5	**43.**	312
44.	329	**45.**	356 R27
46.	1,181 R4	**47.**	4,751

Page 154

48.	2 R163	**49.**	1 R69
50.	1 R70	**51.**	1 R332
52.	7 R177	**53.**	1 R726
54.	36 R26	**55.**	51
56.	10,994 R1	**57.**	20 R120
58.	50	**59.**	60
60.	190	**61.**	250

Unit 6

Page 155

1. 47

2. 72

Line up the digits. Add the ones. $5 + 7 = 12$.
Carry 1 ten. Add the tens. $1 + 2 + 4 = 7$

$$\begin{array}{r} 1 \\ 25 \\ + 47 \\ \hline 72 \end{array}$$

3.	251	**4.**	150
5.	885	**6.**	2,100
7.	7,305	**8.**	10,993
9.	41,100	**10.**	112,037

Page 156

11. 53

12. 10

Line up the digits. Subtract the ones.
$4 - 4 = 0$. Subtract the tens. $8 - 7 = 1$.

$$\begin{array}{r} 84 \\ - 74 \\ \hline 10 \end{array}$$

13.	47	**14.**	158
15.	79	**16.**	2,488
17.	3,277	**18.**	4,358
19.	5,830	**20.**	10,359
21.	128		

22. 600

Multiply by 0 ones. Multiply by 3 tens.
Add the partial products.

$$\begin{array}{r} 20 \\ \times 30 \\ \hline 00 \\ + 60 \\ \hline 600 \end{array}$$

23.	1,476	**24.**	267
25.	3,240	**26.**	4,140
27.	26,871	**28.**	278,388
29.	128,940	**30.**	206,500
31.	51		

32. 61 R5

Divide. $43 \div 7 = 6$, plus an amount left
over. Multiply. $7 \times 6 = 42$. Subtract. $43 - 42$
$= 1$. Bring down 1. Divide. $12 \div 7 = 1$, plus
an amount left over. Multiply $7 \times 1 = 7$.
Subtract. $12 - 7 = 5$. There is a remainder of
five.

$$\begin{array}{r} 61 \text{ R5} \\ 7\overline{)432} \\ - 42 \\ \hline 12 \\ - 7 \\ \hline 5 \end{array}$$

33.	104	**34.**	70
35.	92		

Page 157

1. 255

$5 + 250 = 255$ milligrams. There were 255
total milligrams of cholesterol in her
breakfast.

2. 150

$75 \times 2 = 150$ milligrams. She will take in
150 milligrams of cholesterol.

3. b, 134

$67 \times 2 = 134$ milligrams of cholesterol in 6
ounces of chicken.

4. c, 8

75 − 67 = 8. There are 8 more milligrams of cholesterol in 3 ounces of beef.

Page 158

1. 72

6 × 12 = 72. She is 72 inches tall.

2. 3,520

2 × 1,760 = 3,520. She rode 3,520 yards.

3. c, 108

9 × 12 = 108. Each board is 108 inches long.

4. b, 3

9 ÷ 3 = 3. Each board is 3 yards long.

Page 159

1. 30

80 − 50 = 30 inches. Thirty more inches of rain fell in July.

2. July 40 × 2 = 80

The rainfall for August was half the rainfall for July.

3. d, 5

10 − 5 = 5. Five more inches of rain fell in April.

4. a, 120

80 + 40 = 120. 120 total inches of rain fell in July and August.

Page 160

1. 3,240

500 + 1,120 + 1,120 + 500 = 3,240. Carolyn jogs 3,240 feet in one round trip.

2. 1,620

500 + 1,120 = 1,620. She walked 1,620 feet.

Page 161

3. 2,000 feet

500 + 500 + 500 + 500 = 2,000. She walked 2,000 feet in all.

4. 3,620 feet

1,120 + 1,000 + 500 + 500 + 500 = 3,620. Brad drove a total of 3,620 feet.

5. B, C, H, G

6. 3,000 feet

500 + 500 + 500 + 500 + 500 + 500 = 3,000 feet (or 500 × 6 = 3,000)

7. No

500 + 500 + 500 + 500 + 500 + 500 = 3,000 (or 500 × 6 = 3,000). The perimeter of the park is the same distance Bob walked.

8. 5,500 feet

500 × 11 = 5,500. The shortest path would be 5,500 feet.

Page 162

1. $40

2 × $100 = $200. 2 × $10 = $20. $20 × 12 = $240. $240 − $200 = $40. They will save $40.

2. $168

4 × $7 = $28. $28 × 6 = $168. The grandparents spend $168 for the 6 concerts.

Page 163

3. 302

1,595 + 100 + 530 = 2,225. 2,527 − 2,225 = 302. They sold 302 senior citizens' tickets.

4. 2,750

100 + 150 = 250. 3,000 − 250 = 2,750. 2,750 seats are not reserved.

5. $22,920

1,926 × $10 = $19,260. 732 × $5 = $3,660 $19,260 + $3,660 = $22,920. The concert hall made $22,920.

6. $5,320

342 + 418 = 760. 760 × $7 = $5,320. The concert hall made $5,320.

7. $6,315

1,263 × $5 = $6,315. 1,263 × $10 = $12,630 $12,630 − $6,315 = $6,315. They would have made $6,315 more.

8. 175

1,900 + 925 = 2,825. 3,000 − 2,825 = 175. They sold 175 senior citizens' tickets.

Page 164

1. $11,083

$10,200 + $11,600 + $9,072 + $13,460 = $44,332. $44,332 ÷ 4 = $11,083. The average of their expenses was $11,083.

2. $82

$80 + $75 + $91 = $246. $246 ÷ 3 = $82. The average was $82.

Page 165

3. 9

10 + 8 + 9 + 11 + 7 = 45. 45 ÷ 5 = 9. He worked an average of 9 hours each day.

4. 41

40 + 42 + 38 + 44 = 164. 164 ÷ 4 = 41. He worked an average of 41 hours each week.

5. $33

$28 + $30 + $29 + $33 + $45 = $165. $165 ÷ 5 = $33. She made an average of $33 each day.

6. $153

$165 + $155 + $130 + $162 = $612. $612 ÷ 4 = $153. She made an average of $153 each week.

7. 255

300 + 240 + 240 + 200 + 270 + 280 = 1,530. 1,530 ÷ 6 = 255. His average score was 255.

8. 255

300 + 200 + 250 + 250 + 280 + 250 = 1,530. 1,530 ÷ 6 = 255. Her average score was 255.

Skills Inventory

Page 166

1. thirty-two
2. two hundred forty-six
3. two thousand, three hundred sixteen
4. 50 > 40
5. 15 < 25   **6.** 31 > 13
7. 100 > 69   **8.** 20
9. 2   **10.** 200
11. 5,000   **12.** 90
13. 480   **14.** 3,270
15. 300   **16.** 6,400
17. 45,500   **18.** 5,000
19. 72,000   **20.** 726,000
21. 14   **22.** 18
23. 71   **24.** 69
25. 693   **26.** 3,698
27. 154   **28.** 167
29. 1,278   **30.** 5,294
31. 372,409   **32.** 3,661,515

Page 167

33. 8   **34.** 8
35. 11   **36.** 83
37. 5   **38.** 361
39. 5,134   **40.** 3,332
41. 17   **42.** 64
43. 24   **44.** 25
45. 177   **46.** 559
47. 2,002   **48.** 11,095
49. 153,457   **50.** 42
51. 93   **52.** 2,004
53. 18,090   **54.** 372
55. 18,060   **56.** 48,510
57. 370,829   **58.** 710,000
59. 3,427,000   **60.** 192
61. 1,401   **62.** 29,456
63. 513   **64.** 479,120
65. 84,534   **66.** 214,200
67. 2,425,460

Page 168

68. 2   **69.** 8
70. 81   **71.** 61
72. 6 R1   **73.** 7 R2
74. 73 R1   **75.** 218
76. 251 R1   **77.** 5,240
78. 1 R11   **79.** 9 R34
80. 99   **81.** 808 R20
82. 8 R416   **83.** 102 R14

KEY OPERATION WORDS

Word problems often contain clue words that help you solve the problem. These words tell you whether you need to add, subtract, multiply, or divide. The lists of words below will help you decide which operation to use when solving word problems.

Addition

add
all together
and
both
combined
in all
increase
more
plus
sum
total

Subtraction

change (money)
decrease
difference
left
less than
more than
reduce
remain or remaining
smaller, larger, farther, nearer, and so on

Multiplication

in all
of
multiply
product
times (as much)
total
twice
whole

Division

average
cut
divide
each
equal pieces
every
one
split

TABLE OF MEASUREMENTS

Time

60 seconds = 1 minute
60 minutes = 1 hour
24 hours = 1 day
7 days = 1 week
52 weeks = 1 year
12 months = 1 year
365 days = 1 year

Weight

16 ounces = 1 pound
2,000 pounds = 1 ton

Length

12 inches = 1 foot
36 inches = 1 yard
3 feet = 1 yard
5,280 feet = 1 mile
1,760 yards = 1 mile

Capacity

8 ounces = 1 cup
2 cups = 1 pint
4 cups = 1 quart
2 pints = 1 quart
4 quarts = 1 gallon
8 pints = 1 gallon
16 cups = 1 gallon

WHAT'S NEXT?

You have just finished *Whole Numbers*, the first book in the Steck-Vaughn series, *Math Matters for Adults*.

The next book, *Fractions*, provides easy-to-follow steps and practice in working with fractions. The book begins with a Skills Inventory test that you can use to discover your strengths and weaknesses, and it ends with a second Skills Inventory test that you can use to measure the progress you've made.

In *Fractions* you will learn to solve real-life problems using fractions. There are many uses for fractions in your life. For example, when you share a pizza with 3 other people, each person gets a fraction of the whole pizza. Or when you figure out how much overtime you've worked, you measure your time in fractions of hours.

How many other situations can you think of in which you might use fractions? List them on the lines below.

_____

_____

_____